Fire Protection for Commercial Facilities

Fire Protection for Commercial Facilities

Mark Bromann, SET, CFPS

CRC Press
Taylor & Francis Group
Boca Raton London New York

CRC Press is an imprint of the
Taylor & Francis Group, an **informa** business

CRC Press
Taylor & Francis Group
6000 Broken Sound Parkway NW, Suite 300
Boca Raton, FL 33487-2742

First issued in paperback 2019

© 2011 by Taylor and Francis Group, LLC
CRC Press is an imprint of Taylor & Francis Group, an Informa business

No claim to original U.S. Government works

ISBN-13: 978-1-4398-2720-8 (hbk)
ISBN-13: 978-0-367-86476-7 (pbk)

Library of Congress Cataloging-in-Publication Data

Bromann, Mark.
    Fire protection for commercial facilities / Mark Bromann.
        p. cm.
    Includes bibliographical references and index.
    ISBN 978-1-4398-2720-8 (hardback)
    1. Commercial buildings--Fires and fire prevention. I. Title.

TH9445.S8B76 2010
693.8'2--dc22   2010038201

Visit the Taylor & Francis Web site at
http://www.taylorandfrancis.com

and the CRC Press Web site at
http://www.crcpress.com

Contents

Preface

Have you ever sat in on a meeting when the issue of fire protection was raised, and suddenly felt that you were watching a game from the sideline? Believe me, you are not alone if you just answered "yes" to that question. The fact is that when confronted with a fire protection problem, building management is desperately short on information and know-how in this critical component of protection for its facility. The aim of this text is to touch on the many subfields of fire protection engineering as they pertain to the regimen of the facility manager. The separate issues of fire prevention and life safety presented in this book are more easily digested one topic or subject at a time, and are best suited as a collection included in a single handy reference. None of this material is very hard to grasp. However, there remains the fact that there is simply a lot of basic technical information to absorb before one is sufficiently educated in the administration and application of fire safety.

The intended audience of this book includes not only facility managers and their staff but also fire safety professionals, educational institutions, building engineers, seminar presenters, insurance companies, and consulting engineers. Rather than being written as a running commentary, the text has been formatted to provide information separately by topic for the reader's reference, reflecting current time-proven fire protection practices. Although no book written on fire protection can hope to be completely comprehensive, information is offered here in hopes of augmenting the building or plant manager's broad range of expertise, keeping him or her abreast of the divergent subfields of fire prevention. This knowledge will also aid in decision-making regarding what can be effectively controlled in-house, and what should be contracted out, should the delegation of that work relieve the workload burden of the in-house staff.

To put a number on the daily concerns and duties of the facility manager would be an endless pursuit. He or she is answerable to an unending litany of building needs and is often held accountable to follow-through on numerous nondescript demands from tenants and management. As the day-to-day responsibilities and deadline affairs mount, the tendency is to allow an interest in something "seemingly" less critical as fire protection, to waffle and fall by the wayside. This is a dilemma that the wise manager must overcome through his or her own volition. Fire protection and life safety are items that must be ministered to and properly maintained. They are an integral part of the facility manager's repertoire of expertise. This book will help the reader glean the related facts necessary in order to obtain a good command of fire protection and fire prevention basics.

In the last five years over 25,000 people died in the United States as a result of fires. Half of those killed were elderly, handicapped, intoxicated, or children. Tempering these cold facts is the knowledge that in sprinklered properties, 83% of all fires are controlled by less than three activating sprinkler heads where the system is wet-pipe. But to many, the benefits of fire protection are far from obvious. Further compounding this situation is a major factor that plays a role in keeping these tragic fire statistics as large as they have been—public apathy. We can witness this reality today in an extreme form when, with the emerging trend of local codes attempting to mandate relatively inexpensive residential sprinklers, the knee-jerk reaction of many home-building organizations is nothing short of hostile. Residential sprinklers are but one example of newer technology in the fire protection industry that has displayed an outstanding track record. But when considered alongside other advancements made in this electronic age, today's improved fire protection measures must also be viewed as examples of new technologies that can easily be utilized, so that all facility occupants can enjoy a safe living or working environment.

Public protection from fire is one of the many responsibilities of building management. If the attitude of any professional continues to steer him or her toward ongoing education in a wide array of fields related to his or her own, he or she can only improve in his or her job. It is the aim of this text to assist those who wish to nourish their own proclivity toward the accumulation of knowledge. This activity is proactive and of benefit to many. Knowledge of the perils of fire is the engine that drives fire protection professionals. Although no one may come right out and say it, your interest in fire protection and life safety is appreciated.

The aim of this text is not to qualify readers as certified safety professionals, but rather to expose them to the wide array of fire protection engineering applications encountered during typical facility operations so that they are well versed and informed. Every facility manager dreams of the day when absolutely nothing goes wrong, the week where no new unforeseen problems occur, and nothing gone awry is brought to his or her attention to be reckoned with immediately (if not sooner!). But this is just not realistic. Sooner or later, problems will occur and these may well require some kernel of expertise involving fire protection. Being prepared is half the battle—if not more.

Author

Mark Bromann, SET, CFPS, is a 1977 graduate of Valparaiso University. He is the author of a college-level textbook, *The Design and Layout of Fire Sprinkler Systems,* and his career in fire protection has spanned five decades. Mark's fire protection design experience includes two structures (the DeZwaan Windmill and the AON Building) that appear photographed in the *World Book Encyclopedia.* He is the recipient of an Illinois Engineering Design award (1995) in retrofit fire sprinkler system design work for a high-rise structure in Chicago.

Mark's background includes experience in fire alarm system design, property inspections, technical assistance for attorneys, hydraulic design of all types of fire suppression systems, fire pump testing, sprinkler system plan review, expert witness testimony, and educational services. Mark has also served as an instructor in the Fire Science Technology Department at Harper College in Palatine, Illinois.

Mark has lectured at seminars in the Midwest and on the East Coast on the subjects of residential fire sprinklers, backflow prevention, sprinkler system design, and commercial property inspections. He has testified before various committees and agencies on matters relating to fire protection for business owners and special interest groups, and has served on municipal building boards for matters relative to code development. He is currently president of Rally Fire Protection Services in Wheaton, Illinois.

1

Fire Prevention Basics

It has been stated that you cannot control nature: in the end it controls you. This veiled warning has held true to form until fairly recently in human history. We now understand how fire losses can be mitigated and in many cases prevented from their onset. Today, the services of fire protection engineers are vigorously sought out by those in the business community who desire overall and ongoing fire protection. The principle of all fire prevention can be succinctly summed up by the word "precaution." A precaution is any measure taken beforehand to ward off unwanted events and their resultant destruction. These precautions, or fire protection practices, are globally accepted as part of the norm with respect to the construction of new buildings. In addition, "retrofit" fire protection installations within existing buildings constructed prior to the existence of modern building codes are projects that take place frequently. Those who vehemently oppose the expenditures and efforts made for fire protection measures are mired in an antiquated public indifference, a fossil stuck in the soil of ignorance and shortsightedness. A fellowship is shared on a professional level by those who know better.

Consider this: more than 150 workplace fires occur every day in the United States. Contrary to "conventional wisdom," the United States is home to the highest rate of fire-related deaths in the world. With over 100,000 cases of deliberately set fires annually, the United States also holds the distinction of being the world leader in arson incident frequency. A recent three-year study puts the annual fire-related death rate in the United States at over 19 per million. The rate in Canada and Japan is also high: at 15 per million. The corresponding figures are only 12 per million in France, about 10 per million in Germany and Australia, 8 per million in Spain, and less than 6 per million in Switzerland. These statistics are pointed out, not to give cause for a manhunt to find the villain in the fire equation of any particular country, but rather to illustrate the necessity and urgency to adopt specific measures for fire prevention. One of the major objectives of all building professionals in industrialized nations has been the widespread adoption of common-sense fire safety strategies, the modernization of codes, and the overcoming of barriers and obstacles to the acceptance of basic fire protection methods and practices at every level of commerce. And the practices adopted need to be defined in concrete terms, not just vague platitudes.

Any fire that occurs in a commercial setting has far-reaching consequences that affect not just the business itself, but the community as a
whole. Should facility operations experience a shutdown of even a few short
weeks, the local economy is weakened, the lives of those who depend upon
the business itself will be significantly disrupted, company records may be
destroyed, and the temporary suspension of business and production may
force major clientele to look elsewhere, giving the competition the upper
hand. Risk is built into every commercial property. To combat the ever-
present threat of fire, there exists a need for a firm of any size to place a
competent individual (or several employees) in a position responsible for
the development of a fire prevention program. Although it is often difficult to measure just how successfully such a program is faring, what the
fire safety professional cannot disregard are three basic arenas of potential
fire loss. These are: factors that could contribute to the ignition of a fire,
those that would contribute to the growth and spread of fire, and those that
provide assistance to fire extinguishment. Although these considerations
vary by facility type, the primary objective of a comprehensive fire safety
program should ensure that all is being done to prevent a fire outbreak by
heightening fire prevention awareness in the workplace, recognizing hazardous conditions, and consistently maintaining records that establish the
testing and inspection procedures specific to the occupancy and structure
being protected. The attention focused on these goals will directly correlate
to a diminished potential for fire loss.

Factors Regarding Potential Ignition (Prevention)

When we look at what may contribute to the possibility of a fire starting
inside a building, we must first identify all fire hazards, including those items
on the premises that may be highly flammable or volatile. Second, we must
establish a regular routine with respect to equipment maintenance. Third,
an employee smoking policy must be firmly established, with a lowered tol-
erance toward offenders. Fourth, employees must be reminded at regular
intervals that fire safety precautions are not to be ignored. The observation
of posted rules for fire safety will enhance control over manufacturing, storage, and processing procedures to a degree that will set perhaps a higher
standard of safety than could a simple fire drill. Finally, the building itself
must be secured at all times for no other reason than to prevent any attempt
against the possibility of arson. In addition, building security must prevent
unauthorized access to life safety systems, alarm panels, and active fire protection equipment.

Factors Regarding Potential Fire Spread (Maintenance Measures)

The key ingredients to the prevention of fire growth within a facility include the integrity of fire-safe building construction, a regimen of general housekeeping procedures, removal of combustible waste from industrial processes, and limiting of the storage height of all inventory. The latter advisory is especially essential when the stored products are of a highly flammable nature. Housekeeping procedures should not overlook upkeep of electrical appliances, the correct usage of electrical and heating systems, checking for corrosion of motor parts, and the periodic checking for leakage of lubricants. Any industrial process that creates flammable dust, or entails cutting, welding, or painting, must be monitored on a regular basis.

Factors That Exist for Fire Control and Emergency Training

Fire control components include in-place fire suppression systems such as automatic fire sprinklers, heat detection systems, manual pull stations, and smoke alarm notification systems. Related concerns include employee training, the existence and proper maintenance of safe exits, signage to provide information and instructions for safety, and the means for known hazard separation within the structure. To preserve occupant safety, maintenance requirements are necessary for all amenities and safety features. When successfully implemented, these degrees of prevention constitute the best insurance policy. Every step counts. What assists fire control to the greatest extent is a properly scheduled performance assessment of these deterrents to the ravages of fire. Finally, any major building or occupancy changes within the industrial workplace should initiate a re-evaluation and an examination of alterations that may affect or compromise the original intent of the fire protection plan.

The fire problem is a reality that hasn't varied much since the first fire from two sticks being rubbed together spread out of control. The reality is that fires happen. If fire grows sufficiently and gains momentum, it can wreak total havoc and destroy everything in its path. It is well documented that some structural fires have grown to such large proportions that the responding fire departments were left with no other tactical option but to "let the fire burn itself out." Most adults are acquainted with these facts. The information in the following paragraphs is included with the intent that the more a person knows about the phenomenon of fire development, the better equipped he or she will be with the strategies that can be implemented to prevent, intervene, or halt the fire process.

For our use, fire is defined as any instance of uncontrolled burning. Fire, what the Hawaiians refer to as *keahi*, originates with a fast chemical reaction that involves the burning, or rapid oxidation, of a solid, liquid, or gaseous material. All three elements of fire: oxygen, heat, and fuel, must be present in the proper proportion for fire to start and sustain itself. The air we breathe is about one-fifth oxygen. Heat is the energy necessary to increase the fuel temperature to a sufficient point where ignition can occur. Fuel is anything that will burn. Chemical reactions are changes that occur when forms of energy such as heat are applied to substances, which in turn form new substances. Under the right heat conditions, fire is a chemical chain reaction between the substances oxygen and combustible fuel. The result, combustion, produces flame, bright light, and "products" well known to us as ash, gases, and other compounds.

The best way to visualize combustion is to understand that nothing actually burns in solid form. The surface of the solid must first be converted into a gas (or gases). When someone lights a newspaper on fire, what is really happening is that the *heat* of the match drives combustible gases from the paper into the air, which causes those gases to react with the oxygen. The subsequent gas to gas burning leads to an eventual vaporization of the newspaper. The entire time, the match flame is burning particles of gas.

The chemical reaction takes place and moves very rapidly across a thin illuminated cone called the flame front. The key ingredients contained within most substances that burn are (hydrocarbons) hydrogen and carbon. When combined with oxygen they produce both water and carbon dioxide. These and other gaseous vapors produced by the fire are very hot, creating more vapors from adjacent fuels and incrementally more ignition. This leads to the production of more and more burnable vapors. If not held in check, the process simply continues until all of the fuel is burned out. Wherever a fire starts, the outcome is predictable and can be quantified. It is known that our sun will run out of hydrogen and burn out in approximately five billion years.

As one more example, suppose a pile of oily rags is left on the floor or in a plastic trash bin, and later starts on fire. What has physically happened is that the oxygen from the air slowly united with the oil contained in the rags, a process called oxidation. And when enough heat accumulated (as the rate of oxidation increased), the rags caught on fire. The subsequent burning produced additional heat, which caused the oxygen to unite with other substances much more rapidly. The oxidation led to combustion in this case because the ignition temperature of the soiled rags was relatively low. Likewise, the ignition temperature of gasoline is quite low (and gasoline will vaporize at temperatures as low as −45°F). A structural steel I-beam combines with oxygen at a very slow rate; its ignition temperature is very high. Some substances have already combined with oxygen as much as possible. One is the mineral asbestos. Other examples include stone and sand. Essentially, they do not have an ignition temperature. Oxidation does not always lead to combustion, as in the case of the rusting of iron or the browning of a half-eaten apple.

As previously outlined, there are numerous precautionary methods by which to battle fire before it starts. One of the most important methodologies is accomplished by simply identifying the potential hazards. Certainly, the highly flammable materials must be somehow cut off and separated from other items in the building in a safe and conservative fashion. Arrangements of stored products must be wisely planned and also limited (in terms of height) as much as realistically possible. A building should have separate smoking areas. If you look around a modern structure, fixed fire protection is readily visible: manual pull stations, a mixed array of alarms, fire extinguishers, fire sprinkler or suppression systems, smoke detectors, and hose stations are common examples. Automatic fire sprinkler systems are unquestionably the most effective, reliable, and efficient means for fire protection available today. Not counting New Hampshire, Delaware, North Dakota, and South Dakota, all of the remaining 46 states have experienced at least one major fire in the last century in which more than 10 people perished. In 1998 alone, 35 catastrophic U.S. fires killed 191 people. Contrast that with the fact that there has never been multiple loss of life inside a fully sprinklered building.

Fire alarm systems consist of any number of alarm-initiating devices connected to an electrical network that will transmit alarm signals to the appropriate fire service. Whereas protection of property is the primary focus of sprinkler systems, protection of life can be identified as the primary goal of fire alarms. If, for example, the occupants of a building can't move freely on their own to escape a fire, a supervising station connection can summon aid from the public fire department or from a private fire brigade. Such a connection significantly enhances the life safety goal. At a facility with complex property protection issues, the fire alarm system might be used as a tool to help manage overall fire protection. In this case, the fire alarm system can actually oversee the operational readiness of all the other active fire protection systems.[1]

> A current trend in real estate development has given birth to many new or renovated multi-occupancy buildings. A typical example of this is a bank or office building containing one or more upper floors of apartments or condominiums. Knowing how often fire strikes in the middle of the night, I cannot overestimate the magnitude of life safety concerns for a building that will have sleeping occupants.

If our ancestors had one overriding nighttime fear, it was not witches or even cutthroats. It was fire. Blazes were common in congested cities: houses, with wood frames and thatched roofs, ignited easily. And open flames flickered everywhere, especially at night … also, householders commonly complained about servants forgetting to bank fires or snuff out candles. Victims of nighttime street attacks often yelled not Help! or Murder!—but Fire! That cry was most apt to bring out the neighborhood.[2]

It is true today that residential fires, for a host of reasons, comprise the core of the fire-related death problem in any country, about 80% of it. If it comes to be that any part of a commercial building is classified as residential, all measures taken for fire protection must be stepped up according to code. The inclusion of fire sprinklers should be a given and again, planning is essential. See Figure 1.1.

Other basics common to fire protection technology are ongoing and people-related: exit drills, education and training, safety programs, inspections, equipment testing, and building maintenance to name a few. When a building is constructed, architectural planning is reviewed to make sure that the fire department has easy access to the structure, building units are "compartmentalized" for fire stoppage, and fire-resistive construction materials are used. Requirements called for by building and fire prevention codes are extensive. They range from structural fireproofing and oversized corridors/stairways to separations between buildings and fire-resistant roof coverings; from the control of fires, explosions and smoke in heating and air handling systems to numerous systems and devices for the detection, alarm and suppression of fire; and from one large volume on safe electrical systems to many volumes on inspections, testing, maintenance, training and drills needed to keep everything working and everyone properly informed.[3]

> What will also mitigate fire-related problems and injuries is a working fire safety program. Active participation by both building occupants and property management is necessary for any such program to achieve the goal of complete fire safety. The first step in developing a relevant safety program for any building is to select a building manager responsible for its creation. He must be keenly aware of what hazards and abnormalities are particular to his building, and tailor his program in accordance with those particulars. He must also realize that any amount of prevention and planning that is adopted is infinitely better than none at all. Just observing the simplest of fire precautions will lessen the chance of fire.

The most difficult part of drafting an outline for a long-term fire safety program is the basic exercise of getting started. Once you begin filling in a blank sheet of paper, the ideas and concepts will come. The approach should be community-based, stressing awareness and participation. The plan should articulate actions that include or enhance the fire safety requirements noted in the model building codes. Finally, the plan should not stress any strategic concepts that detract from the requirement or aim of complete automatic fire sprinkler protection.

To review, the fire safety program outline should contain at least three key headings: Prevention, Maintenance Measures, and Emergency Training. All notes (goals) listed beneath these headings must be specifically defined. For example, beneath the "Prevention" heading, we obviously desire that the occupants conscientiously observe all fire safety rules. But pertinent informative procedures are what we're after. For example, Item #1 could

FIGURE 1.1
A firefighter checks for lingering smoke and flames following the extinguishment of an overnight fire. (See color insert following p. 52.)

consist of the elimination of fire hazards such as space heaters, long extension cords, multiple plugs, and open flammables. Combustible liquids
should be stored in sealed containers in approved areas. Item #2 could
address the means by which the property can be secured to ward against
possible attempts at arson. Item #3 could list an applicable housekeeping
policy: keeping combustibles a safe distance away from heat sources, providing a three-foot clearance between any items and all electrical service
equipment, and keeping small appliances disconnected from the power
source when not in use. Item #4 could note a list of areas in need of repair:
holes in walls or vent piping, openings in ceilings or flues, or faulty closing
devices on fire doors. Open wiring must be enclosed, and so on. This is a
listing of items that need someone's attention, and must be detailed with
consideration given to the specific manufacturing, storage, and processing
that goes on within the building. Remember that what can burn is what is
brought into a building.

Maintenance Measures are steps taken to increase the level of occupant safety. Exit facilities are to be maintained in a clean and open condition. Access to fire protection components by authorized personnel
must be ensured. The visibility of warning and exit signage should be
confirmed. Exit routes must be posted. Records and reports should be
kept with regard to equipment inspection as well as maintenance activity. And the entire facility must be assessed periodically to keep current
with code changes. Fire alarm systems typically monitor the operational
status of other important fire protection systems, such as automatic sprinkler systems. If the owner chooses not to maintain the fire alarm system,
then two important fire protection systems that provide life safety might
become adversely affected. ... It is important that local enforcement officials remind owners of their code-required obligation to maintain their fire
alarm systems and ensure that these systems remain operationally reliable.
No matter how *bad* the economy may become, cutting costs by failing to
maintain critical protective systems will have serious consequences. No
amount of savings can ever make up for a potential loss of life or destruction of property because an owner has chosen to not maintain a protective
system.[4]

Emergency Training includes emergency action plans. This pertains
to fire drills and alternative measures for evacuation in the event of
unwanted shutdown of fire pumps, electrical service, or blocked paths of
egress. Depending on the size of the building, any number of supervisory staff members must be designated for assistance. Their responsibilities must be written down so that they realize what is expected of them.
Because immediate action is essential in an emergency, they need to know
what to do in terms of the simplest behaviors. This is extremely important
for exiting purposes. For situations where all hell breaks loose, there is a

need for individuals trained in proper emergency behavior. Rule number one is not to panic. In a multiple-story building, never use the elevator. The closest stairway is your primary escape route. There are several other tangible basics:

- The best air is close to the ground, so stay low.
- Always proceed down the stairs, never go up.
- To help breathe, cover your nose and mouth with a damp cloth if possible.
- Never open a closed door without feeling it first with the back of your hand.
- If trapped, call the fire department first and then try to ventilate the room.
- The last one out of a room should close the door but not lock it.
- Once outside, report to a predetermined area for a head count.

It is prudent to maintain an updated emergency action plan for staff members to distribute. It should include primary and secondary escape routes for all personnel. Any successful fire safety program needs a group of people trained in the ABCs of safe procedures and a commitment to active intelligent response in the event of fire. Grouped as a committee, they must aspire to define objectives, set goals, assign responsibility, and manage the development of the program.

The reader should be enlightened to the fact that the shock of just one visit to a hospital's burn ward would accomplish what one reading of this book would do toward the realization of the necessity of fire protection in today's world. If we live in a civilized society at all, we must possess the civility and responsibility to protect all fellow men and women from the awful consequences that the threat of fire presents. Nearly every community nationwide is served by an organized fire department, yet this alone cannot temper the alarmingly high fire statistics reported every year. Over 2.4 million fires occur annually in the United States, which account for tragedy, waste, and hundreds of thousands of injuries. Think of this when you initiate a move to implement fire protection and safety planning. A member of a popular rock group recently stated, "We're nothing without our fans." Sad to say, but we in the fire protection community barely have any "fans." Public awareness and education campaigns carry with them very noble goals, but they travel on a constant collision path with problems such as public apathy and concerns about economic impact. When they do succeed, the awareness tends to dissipate over time. Persistence is the key. Get the word out, and lead by example.

Endnotes

1. Dean K. Wilson, "Supervising Stations," *NFPA Journal,* May/June 1999, p. 130.
2. Joyce and Richard Wolkomir, "When Bandogs Howle and Spirits Walk," *Smithsonian,* January 2001, p. 43.
3. Bert M. Cohn, "Life Safety Goals for Building Codes," *Professional Safety,* April, 1993, p. 16.
4. Wayne Moore, "No Excuses," *NFPA Journal,* September/October 2009, p. 38.

2

Hazardous Commodities and Conditions

News stories pertaining to fires inevitably contain a phrase similar to "When the flames subsided ..." Everyone knows what happened at that juncture: the fuel had run out because a fire had just consumed every bit of it, and there was nothing left to burn. A familiar saying known to Navy and USAF fighter pilots goes like this, "The only time you have too much fuel is when you're on fire." This clever adage is also sound advice for the facility manager of an industrial or storage warehouse. The analogy is that inside a facility without fuel, fire cannot burn. With an abundance of fuel, the potential exists for a fire to rage uncontrollably. The chance for a fire to begin, how quickly the fire spreads, and how destructive the fire can ultimately become, depend upon an informed and close examination of the fuel itself.

Suppose that your place of business experiences a fire. It is likely what burned in this scenario were either solids or flammable liquids. In the aftermath of an actual fire, various business personnel will be asked a number of questions, most of which have to do with "property," that is, the commodities that were contained in their place of business. An accounting will be requested, a succinct catalogued recording of everyday items that the firm uses in any quantity that may ignite when fire or sufficient heat is present. In a perfect world, an inventory of this type should have been formulated as part of a total preventive fire safety effort. Building engineers must be aware of how toxic, flammable, explosive, reactive, radioactive, or corrosive the building contents are by nature, and in combination with other materials present. For any facility, the most important information that can be supplied to firefighters involves the identification of existing hazardous materials. Interaction with local fire officials also gives you the following to think about in terms of a potential fire:

- What are the potential sources of ignition?
- How fast will the fire grow?
- Will the alarms respond rapidly?
- Where are the optimum paths of egress?
- What effects will the resulting smoke have on life and property?
- How can the fire be controlled?

The conditions under which the Incident Commanders and the Hazardous Materials Control Officers of any organized firefighting team attempt to control and mitigate incidents involving hazardous materials are often unpredictable and always very stressful. Because of these difficulties, the effectiveness of their strategies and tactical decisions are highly dependent on the integrity, quality, and accuracy of the information provided to them.

Fuel for the Fire: Is It "Hot" or Is It Not?

When we concern ourselves with the fire hazards of solids and liquids (and dusts and gases) present in any building, we need to investigate the volatility of these materials because their fire potential differs considerably due to variations in burning characteristics. The most important items to spot, of course, are those that represent the greatest threat. To measure and identify those that are the most menacing to your business, the following three factors must be considered: the flash point of the material, its ease of ignition, and (especially in the case of compressed gases) its ignitability limits. When we concern ourselves with the fire potential of solids, liquids, gases, and combustible dusts, we need to investigate thoroughly the volatility of these materials.

Solids

The volatility of any solid, and that includes the structural building materials (wood, steel, concrete), can always be tempered to some extent by flame-resistive treatments or fireproofing. What makes a solid most susceptible to combustion is the amount and degree of heat that is applied to its surface over a certain duration of time. With proper flame-resistant treatments, flame spread can be effectively diminished even if ignition has already occurred, thereby depreciating heat release rates. This is especially true when the material is wood. When the potential combustible is a stored commodity as opposed to a part of the building structure, there is limited "fireproofing" potential. Realistically we cannot fireproof everything in the warehouse; this is the basic reason for installing automatic fire suppression systems. Far from being a stopgap measure, fire sprinkler systems are designed to control or extinguish fire by a design in congruence with the existing commodity hazard and conditions and arrangements of storage. This sprinkler system design and its accompanying water supply are bolstered to a recommended minimum degree to protect products with higher burn rates, such as rubber tires, roll paper, certain plastics, and upholstery containing plastic foam.

Most plastics when stacked in storage warehouses will burn at the same rate as stored cardboard boxes, fabrics, and wood. What burns much more intensely are foam rubber and foam plastic, which will also produce a virtual cloud of smoke and toxic gas. When used as wall or ceiling insulation, foam plastics are covered and concealed by gypsum wallboard which constitutes a fire-resistant barrier. This effective fire solution is not an option when dealing with storage racks filled with mattress pads or furniture cushions. Buildings that warehouse these items, or highly flammable products such as tires, plastic packing, plastic insulation, roll paper, furs, or cellulose nitrate film, will require the installation of smoke and heat vents. As with the increased capability of the fire sprinkler system, the number and size of the vents must be specifically tailored to the anticipated storage height and the quantity of the commodity that is present. You cannot play poker with this: it is a necessary life safety measure as well as an aid to the responding firefighting team.

Specifics for the classification of commodities are found in NFPA (National Fire Protection Association) pamphlet #13. Table A.5.6.3 in this document offers an alphabetized commodity listing to identify the corresponding hazard group category. Products to be stored higher than 12 feet are designated by one of four classifications: Class I being the least volatile, Class IV being among the highest of hazards, and are further established as follows:

Class I: encompasses essentially noncombustible products in paper or cardboard cartons or wrappings, with or without wooden pallets.

Class II: is composed of Class I products in wooden or multilayer cardboard containers, with or without pallets.

Class III: contains wood, paper, or natural cloth products or Group C plastics, with or without pallets. A limited amount of Group A or B plastics may also be included.

Class IV: is composed of any of the above, with an appreciable amount of Group A or B plastics in ordinary corrugated cartons or with Group A or B plastic packing, with or without pallets.

Intense, fast-developing fires can result where large quantities of plastics or rubber are stored. Those commodities are classified into groups (A, B, or C) with Group A representing the most difficult of the potential fires to combat. Examples of specific Group A commodities are listed in Section 5.6.4.1 of NFPA #13 (2010 edition). The next tables in succession (see below) note the appropriate classification of Group B and C plastics and rubber by material composition. For example, ethylene fluoroplastics and silicone rubber fall into Group B, PVC polyvinyl chloride and polyvinyl fluoride fall into Group C, and polyethylene, polypropylene, and butyl rubber fall under the Group A category.

Group A

Acrylonitrile–butadiene–styrene copolymer (ABS)

Acrylic (polymethyl methacrylate)

Acetal (polyformaldehyde)

Butyl rubber

Ethylene–propylene rubber (EPDM)

Fiberglass-reinforced polyester

Natural rubber (if expanded)

Nitrile rubber (acrylonitrile–butadiene rubber)

Polybutadiene

Polycarbonate

Polyester elastomer

Polyethylene

Polypropylene

Polystyrene

Polyurethane

Polyvinyl chloride (PVC, highly plasticized; e.g., coated fabric, unsupported film)

Styrene acrylonitrile (SAN)

Styrene–butadiene rubber (SRB)

Thermoplastic polyester (PET)

Specific product examples of these commodities are listed in the NFPA #13 appendix.

Group B

Cellulosics (cellulose acetate, cellulose acetate butyrate, ethyl cellulose)

Chloroprene rubber

Fluoroplastics (ECTFE, ETFE, and FEP)

Natural rubber (not expanded)

Nylon 6 and nylon 6/6

Silicone rubber

Group C

Fluoroplastics (PCTFE and PTFE)

Melamine (melamine formaldehyde)

Phenolic

PVC (polyvinyl chloride, flexible; PVCs with plastic content up to 20%)

PVDC (polyvinylidene chloride)

PVDF (polyvinylidene fluoride)

PVF (polyvinyl fluoride)

Urea (urea formaldehyde)

The suppression of any fire involving these solids will be best accomplished with large volumes of water. Water piped from a fire sprinkler system will be cold and it removes heat from a fire. This breaks up one key ingredient of what is commonly referred to as the fire triangle (heat, oxygen, and fuel). Due to its high specific heat capacity, water takes away heat more effectively than almost any other everyday substance. Water-based portable extinguishers (Class A or Class ABC) must be present, and in-place standpipe and sprinkler systems must be specifically designed for each storage arrangement. It is extremely important to recognize that any fire involving plastics or rubber can and will generate a prodigious amount of dense toxic smoke. The entire building must be immediately evacuated and responding firefighters should be forewarned of the respiratory danger.

It is critical to recognize materials on which it is not advisable to use water as an extinguishing agent. This is very important when certain chemicals are involved. For example, the application of water on quantities of aluminum powder, calcium carbide, calcium phosphide, metallic sodium and potassium, quicklime, magnesium powder, and sodium peroxide may pose a dangerous threat to personal occupant safety and will do nothing toward the termination of fire. The chemical properties of all materials must be reliably established. Certain chemicals actually have the ability to oxidize other materials. Specific chemicals that have toxic, combustible, unstable, or reactive properties must be handled within a closed system. The insurance carrier of any business employing the use of these chemicals will certainly provide a long litany of control and protection requirements.

Liquids

The specific *flash point* of any liquid is what is used to differentiate *flammable* from *combustible* liquids. By definition, a flash point is the lowest temperature at which a liquid gives off enough vapors to ignite momentarily in air. In the presence of flammable vapors, rising heat, a spark, or some other source of ignition will be enough to quickly start a blazing fire or in some cases (when in a confined space) an explosion.

A flammable liquid (classified as a Class I liquid) is any substance that is in liquid form at ordinary temperatures and possesses a flash point below 100°F. Gasoline is one common example of a flammable liquid. In U.S. residential homes alone, there are over 5,000 gasoline fires annually resulting in approximately $100 million in direct property damage. Gasoline is extremely volatile and must be stored in small amounts in separate "cutoff" rooms. The smaller the containers, the better they are. Liquid containers of five gallons or less, called safety cans, are virtually leakproof and provide a means of safely transporting the liquid. In addition, the containers:

- Contain automatically closing fill and dispense openings
- Vent excess pressures to ward against vapor explosion
- Prevent external flames from coming into contact with the liquid
- Clearly mark the container contents
- Will bear a U.L. or F.M. approval designation

All flammable liquid containers are to be kept in a remote location, within a metal cabinet or "flammable liquid locker" designed specifically for this purpose. The containers should be marked with one of the following subclassifications:

Class IA: Liquids with flash points below 73°F and boiling points below 100°F

Class IB: Liquids with flash points below 73°F and boiling points above 100°F

Class IC: Liquids with flash points above 73°F and boiling points below 100°F

A combustible liquid has a flash point equal to, or above, 100°F. These are subdivided as follows:

Class II: Liquids having flash points between 100°F and 140°F

Class IIIA: Liquids having flash points between 140°F and 200°F

Class IIIB: Liquids having flash points in excess of 200°F

The reasoning behind such well-defined category delineation is for code compliance and suppression system design. As shown in Table 2.1, hazard control measures are taken to much more stringent levels based on the degree of hazard and the total amount of liquid to be stored and handled.

Most facilities utilize flammable and combustible liquids to some extent in everyday operations in the form of solvents, lubricants, grease, and fuels. Some are contained in aerosol cans, which pose their own special threat. The most fundamental safety measures to be taken consist of keeping the liquids in closed containers as much as possible to minimize their exposure to air. They should also be kept a reasonably safe distance away from electrical and heat-producing equipment. Fixed fire suppression systems must be regularly maintained, annually tested, and designed in accordance with the recommendations noted in NFPA #30, *Flammable and Combustible*

TABLE 2.1

Indoor Unprotected Storage of Liquids in Containers and Portable Tanks

Class	Container Storage		Portable Tank Storage	
	Max. Piling Height (ft)	Maximum Quantity (gal)[a]	Max. Piling Height (ft)	Maximum Quantity (gal)[a]
IA	5	660	—	Not permitted
IB	5	1,375	7	2,000
IC	5	2,750	7	4,000
II	10	8,250	7	11,000
IIIA	15	27,500	7	44,000
IIIB	15	55,000	7	88,000

[a] Refers to overall total quantity in facility, not just in a single pile.

Liquids Code. The best fire extinguishers to mount nearby are either foam or BC-rated dry chemical extinguishers. Other recommended safety measures include:

- Explosion venting of specially designated rooms (especially essential for Class 1A and other unstable liquids
- Restrictions on liquid quantity within the building, and in individual stacks
- An abundance of air movement (mechanical ventilation)
- If possible, the alternate use of nonflammable liquids (two cleaning solvents that are nonflammable at normal temperatures are trichloroethylene and tetrachloroethylene)
- Operations training of personnel
- Safe handling and manufacturing practices

For optimum hazard control, dispensing and storage operations of flammable liquids are best kept in a separate building from the plant, or in a vented "cutoff" room attached to the plant. The guiding principle here is to allow for the escape of flammable vapors. Also, any fire occurring within the cutoff room will be reasonably contained and unable to spread to other areas of the building. Fire hose stations should be in close proximity. The cutoff room must have self-closing doors and a drainage system, usually incorporating a lower (6 in. or greater) floor with scuppers to direct the flow of spilled liquids to areas other than the adjacent facility. It is an unwise practice to keep more volatile liquids in storage than actually necessary for normal plant operations.

Any large fixed equipment such as commercial presses that are typically covered with an oily buildup should be protected by a separate (foam) deluge system. This is especially important when there is a pit beneath because vapors from the oils and lubricants tend to lay in low areas and accumulate at dangerous levels. The following NFPA standards cover requirements and recommendations for specific industrial applications:

NFPA #30: *Flammable and Combustible Liquids Code*

NFPA #30B: *Code for the Manufacture and Storage of Aerosol Products*

NFPA #31: *Standard for the Installation of Oil-Burning Equipment*

NFPA #33: *Standard for Spray Application Using Flammable and Combustible Materials*

NFPA #34: *Standard for Dipping and Coating Processes Using Flammable or Combustible Liquids*

NFPA #35: *Standard for the Manufacture of Organic Coatings*

NFPA #36: *Standard for Solvent Extraction Plants*

NFPA #37: *Standard for the Installation and Use of Stationary Combustion Engines and Gas Turbines*

NFPA #45: *Standard on Fire Protection for Laboratories Using Chemicals*

NFPA #58: *Liquefied Petroleum Code*

NFPA #385: *Standard for Tank Vehicles for Flammable and Combustible Liquids*

NFPA #395: *Standard for the Storage of Flammable and Combustible Liquids at Farms and Isolated Sites*

Finally, there is the matter of the disposal of hazardous waste that contains toxic, explosive, corrosive, or flammable substances. These include (but are not limited to): paints, polishes, lacquers, solvents, automotive fluids, paste waxes, kerosene, turpentine, insecticides, chemicals, petroleum products, inks, varnishes, and stains. Leftover hazardous products are not to be poured into open drains or onto the ground, and must be managed in accordance with approved safety and environmental regulations. Today permanent hazardous waste collection facilities dot the landscape of the United States and other countries, so there is little excuse for lackadaisical activity in this regard. Facilities must also be kept clean and free of combustible rubbish. Safe areas must be designated for the temporary storage of waste products. These products are not to be mixed, and must be removed at regular intervals. If "leftover" hazardous products cannot be used up in the plant, and have to be discarded in some fashion, a portable (wheeled) solvent tank may be used to transport the liquids. This tank must also be vented to avoid an excess of vapor concentration. Although this might entail some degree of risk, it is infinitely more efficient, and perhaps less risky, than taking many small containers to the same destination.

Areas red-flagged as containing known hazards will certainly come into the sphere of influence of an insurance carrier or alert fire inspector.

Activities such as paint or ink mixing, oil refining, chemical processing or plating, varnish dipping, solvent cleaning, and asphalt saturating inherently possess the need for high-hazard technical assistance.[1]

Applicable NFPA standards provide strict safety guidelines. The "any protection is better than nothing" theory makes little sense when hazardous materials are involved. When the uninformed take shortcuts with regard to prevention and control in these situations, they are fighting a battle with outdated weapons. Only in code-compliant buildings of fire-resistive construction, protected by in-place fire suppression systems, that practice scheduled inspection and maintenance procedures, can we assume a reasonable measure of well-being.

The five most common causes of accidental death are falls, automobile collisions, poisoning, drowning, and fire. Fire kills, although flames aren't usually the killer. Burns kill and so does smoke. Blinding smoke spreads faster than the flames and contains (lethal) carbon monoxide. Direct property loss from fire in America is estimated to exceed $9 billion annually; this is a vexing problem and a public concern. When we practice fire prevention we must be primarily concerned with what can burn and how intensely that fuel can burn. Normally examined for fire potential are "solids" and flammable liquids, as they are the typical culprits. But gases and dusts, due to their highly volatile nature and ability to produce mass quantities of heat, must be effectively handled in the workplace. See Figure 2.1.

FIGURE 2.1
Typically, only a stone chimney remains in the aftermath of fire.

Gases

Although there are numerous substances that can exist as either liquids or gases, most exist as a gaseous agent in normal temperatures. A "flammable gas" is one that will burn in our normal atmosphere, and only ignite above a specific "ignition temperature." Gas is lightweight, invisible, and will expand to fill whatever container in which it resides. The model codes will usually refer to "hazardous gases" and classify them by their typical usage:

- *Fuel gases* are burned along with air to generate power, light, or heat for buildings. Examples include butane, propane, and natural gas.
- *Industrial gases* include a wide range of gases, classified by their chemical composition, that are used in industrial applications such as welding, heat treating, refrigeration, metal cutting, water treatment, and chemical processing.
- *Medical gases* are used primarily in hospitals for anesthesia and respiratory therapy. Examples include nitrous oxide and oxygen.

Compressed, Liquefied, and Cryogenic Gases

It is important to know whether a gas is stored in a liquid or gaseous state. A *compressed gas* held in a container is one that exists entirely in a gaseous state. It is ordinarily kept in a high-pressure cylinder built to exacting code specifications, with a diameter that is small in relation to its length. The tank wall thickness is usually proportional to the tank diameter. Portable containers are constructed with a "collar" at the top designed to protect the (manual) container valve and cap. Some containers may be equipped with automatic valves designed to shut off when they sense excessive exterior heat or an excessive gas flow rate. These same containers are used to hold *liquefied gas*, which is contained in a much greater concentration than compressed gas. Liquefied gas exists partially in a liquid state and partially in a gaseous state, and always under pressure. It is stored at normal temperatures, as opposed to *cryogenic gases*, which exist in wholly liquefied states at much lower temperatures. Should cryogenic gas escape in liquid form it can cause frostbite to anyone coming in contact with it. A cryogenic gas cylinder holds the liquefied gas at low to moderate pressures.

Utility (natural) gases are flammable fuel gases, contained in (cast-iron) underground and (steel) overhead piping. They are nontoxic and fairly odorless, but can cause asphyxiation if released in sufficient quantity. When held in a container it is still subject to breakage, but because it does not contain liquid it cannot result in a BLEVE situation (see Table 2.2).

TABLE 2.2

Examples of Specific Gases

Compressed	Liquefied	Cryogenic
Acetylene	Liquefied petroleum (LP-Gas)[a]	Liquefied natural gas (LNG)
Ethylene[a]	Anhydrous ammonia[a]	
Hydrogen[a]	Carbon dioxide	
Oxygen[a]	Chlorine	
	MPS (methyacetylene-propadiene, stabilized)	

[a] May also be in contained as a cryogenic gas.

BLEVEs

A common sight in industrial and commercial facilities is a group of (carbon steel or aluminum) liquefied gas containers stored vertically and chained together, or otherwise bracketed to something structurally solid. It doesn't take a genius to recognize that this manner of storage has been undertaken as a safety precaution against a possible explosion. Explosions involve the rupture of a container, coming as a result of excessive internal pressure. The worst-case scenario in this situation is referred to as BLEVE, or a boiling liquid–expanding vapor explosion. This can occur in any number of different ways:

- Most cylinders come equipped with a pressure relief device that will eliminate the chance of any cylinder rupture due to overpressurization by releasing tiny amounts of gas at a certain preset pressure. But the problem arises when the cylinders contain a highly toxic gas, which is not allowed to be equipped with overpressure-limiting devices.
- Any exposure to fire will gradually weaken the vessel wall, and raise the internal pressure faster than the relief valve can release the gas.
- The overpressure relief device may be rendered ineffective due to corrosion.
- Inadequate ventilation in the immediate area lends itself to fire exposure. The overheating of the metal will lead to a BLEVE situation.
- The overfilling of a liquefied gas container will leave too little space for the vapor, leading to a greatly increased chance of overpressurization, which may eventually overwhelm the feasible container strength of any vessel.

A compressed gas that burns very quickly is acetylene. The concept of chaining together acetylene tanks in a vertical arrangement reduces the potential for BLEVE, because if the vapor from an adjacent tank is burning, then the relief valve on top of the next acetylene tank will pop, causing the gas to escape (any liquid will be at the bottom of the tank), and thus averting a chance for explosion. Some LP-gas tanks are sealed on the exterior with fire-protective coatings. These coatings insulate the vessel from extensive heat and extend the time required for the metal tank to reach the point of rupture, thereby giving the fire service ample time to arrive and break up the escaping gas with a water fog.

Aerosols

For those in the fire protection field, the most frustrating problems come into play when adequate prevention measures have not been implemented. One example concerns the storage of aerosol cans. Aerosols, which are widely used, are found in small (typically 16 oz.) high-strength welded metal cans. The gas may be held at pressures as high as 400 psi, and can produce fires of extremely high temperatures. When ruptured by whatever means, these aerosol containers can become propelled flying rockets of fire, quickly spreading the blaze all over the warehouse and overwhelming the suppression system in the process. Containment is the solution to this quickly escalating fire potential. If specialized protection cannot be achieved with the erection of separate cutoff rooms, the aerosol storage area must (at a minimum) be fenced in up to the roof with approved materials. The magnitude of the problem is such that aerosols must not be stored in a general-purpose warehouse, or alongside any other potentially combustible materials. As with flammable liquids, the suppression systems protecting the storage of aerosols must be augmented and intensified in accordance with the recommendations of the insurance authority or NFPA Pamphlet # 30B. Leaky aerosol containers are a big problem, and personnel should check cans for leakage by submerging them in warm water.

Combustible Dusts

The presence of combustible dusts in any occupancy presents a special problem because not only are these easily ignitable, but the very nature of their existence provides the potential for a flash fire that could spread rapidly beyond its point of origin. The behavior of such a fire is dependent on the size and concentrations of the dust particles, the smaller ones being the most apt to catch fire. "Dusts" come in the form of metal filings, thermoplastic resins, agricultural dusts, sawdust, carbon and coal dust, chemical dusts, and fugitive residues left from the manufacture of pharmaceuticals and vitamins. Even deposits of dusts that form on structural beams represent a considerable danger. Any amount thicker than a butter knife blade is a problem. Despite our extensive knowledge of the possibilities, serious dust explosions continue to occur in large industrial buildings today. The presence of flammable gases where dusts are present severely compounds this sizable safety issue. Ignition sources are identified as being either mechanical (from friction), electrical (electrical faults), or thermal (heat from an external source). Layers of airborne or static dust within a manufacturing facility may be ignited by sparks, hot surfaces, electrical arcs, or open flames, and under certain conditions can produce a dust explosion. To mitigate this possibility from arising, the following good housekeeping practices should be observed:

- Any surface on which dust may accrue must be cleaned at regular intervals.
- Dust-generating equipment and materials should be equipped with (closed) removal systems.
- Cleaning shall not be accomplished by blowing, but by a central vacuum system.
- Hatches, bins, seams, and panels on removal systems must be air-tight.
- Where dust aspirators are needed, removal systems for access panels must be periodically checked.
- Schedule regular dust-removal pickups and stick to that schedule.

Matches are a common ignition source. Any places where dust will accumulate should be designated as a nonsmoking area. Electrical equipment in the immediate area must be periodically checked for safe operations. It's best to confine all dust-handling equipment to explosion-proof rooms. Pieces of equipment that process or handle the dusts should come equipped with explosion vents. This is especially critical where agricultural materials (flour, grain, sugar, etc.) are processed, as those are extremely explosive. One grain elevator will typically be equipped with an array of explosion vents, arranged at separate elevations. Special detectors for explosion protection are available from various manufacturers, in conjunction with small stand-alone surface units that will release a suppressing agent.

The strategies for explosion control can be summed up in the following methods: venting, containment, and suppression. The means for explosion prevention are as follows:

- Eliminate all sources of ignition.
- Operate the process outside flammable limits.
- Include steam, CO_2, or nitrogen as a safeguard to neutralize the process of rapid combustion.

A completely different type of gas (fire-generated, not stored) is that which escapes in a fire scenario. It is the tendency of this escaping gas to form itself in layers, with the hottest gases comprising the top layer. Billowing smoke will contain hot air, combustion particles, and heated gas, which will rise from the fire. Venting at the roof becomes an extremely important fire protection strategy then, not only for the prevention of explosion but also to allow for the quick discharge of gas to the outside. The "thermal layering" of the gases is a veritable blessing, because the lower layers represent a much safer environment for those attempting to escape from a burning building.

Gases and dusts represent a life safety threat wherever they are present. Fire is a strange beast, and its prevention is best accomplished by endorsing a total fire safety effort. Especially when highly flammable fuels are in the mix, the fire prevention methodology employed must be equal to the task.

Endnote

1. Mark Bromann, "Fuel For The Fire," *Fire Protection and Design*, March 2008, pp. 7–11.

3

Selling the Concept

For those of us who work in the fire protection industry on a daily basis, one of the most bewildering things to encounter is someone's basic opposition to the undertaking of measures designed to control and combat fire. We're usually not ready for that. Sometime, somewhere, and often unexpectedly, a critical part of our job will entail marketing the very concept of fire protection. In the best scenario, the need for this sales pitch will be anticipated and can be prepared, as when a speaker must address a municipal building board regarding the necessity of proposed fire sprinkler legislation or a local fire prevention code amendment. But many times the call to elaborate on the practical monetary and life safety benefits of fire protection practices comes out of the blue, and vocal opposition soon takes the form of an unappeased citizen, business representative, or even a building engineer.

Have you ever saved any computer-generated information on a disk? If so, I'm sure that this exercise left you feeling secure in the knowledge that as a result of saving those data, you no longer needed to worry about losing it all if the computer crashed or if someone somehow accidentally deleted that file. Fire protection applications are very similar to using a backup disk, or hitting the "save" key. What you gain is security. The implementation of a fire loss program, or the installation of fixed fire protection, or facilitating the means for education and training are all precursors to the reduction of the risk of fire loss as well as the impact of those potential losses on continuing business operations. The reduction of risk is vital to business survival. The business owners that I have encountered who are most interested in keeping tabs on their firm's methods of maintaining proper fire protection measures are those who have already experienced the ravaging consequences of a fire. They are the ones most aware of the fact that the cost of managing fire potential is much less before a loss has occurred, as opposed to the staggering costs following an actual fire. The concept of fire protection has already been sold to those particular people.

The problem, of course, is selling the fire protection concept to a management team who only views money spent on fire protection as cash out the window, reasoning that it gives no return on its investment. These individuals may be very firm in their belief that loss control is only to be implemented to comply with insurance requirements, compulsory regulations, or to meet corporate policy. Simply put, they are strangers to the long blanket of protection that the benefits of loss control unfurl.

The fundamental nature of the skeptic's question: "How in the world is a fire going to start in here?" represents one of the biggest hurdles that we in the fire protection industry have to clear. Obviously, the person asking this question possesses a mentality, or a concocted belief, that is solidly entrenched in his own mind. He may also take his own local fire department for granted, cynically regarding them as a fraternity of slackers sitting around in the fire station playing poker and watching television. This type of person is depictive of a worst-case scenario, and his views are far removed from even conventional public opinion. The very first thing that he needs to realize is that *fires simply happen.* Does he realize that at this moment, somewhere in the world a building is burning? Or that fire kills over 100 people per week, on the average, in the United States alone? In order to most effectively sell the concept of life safety and fire protection, the critics should first be silenced by statistical documentation, or they may not listen to anything further.

Substantiate Your Position

More than 150 workplace fires occur in the United States every day. Every year, fire is the third leading cause of accidental death in the United States. Year in and year out, these are the facts. There is no need to explain the wide variety of causes responsible for each and every fire. There is no one specific cause, such as smoking, cooking, heating, oily rags, electrical distribution, faulty appliances, lightning, or arson, that can be eliminated in order to rid the world of fire. The stark reality is that fires simply occur, and all of them arrive in completely improbable fashion. They happen during the day and they also strike at night.

Statistics documenting the costs resulting from fire are almost too startling to comprehend. Annual property damage sustained in the United States as a result of fire normally exceeds $8 billion. The losses for storage facilities alone exceed $650 million every year. This is the big picture, and after digesting this information, a business owner should be made cognizant of the compelling fact that, as reported by the National Fire Protection Association, 43% of businesses that experience a large fire never resume business. Considering this, it makes perfect sense that all business concerns need some format for loss control. It is equally clear that if fire prevention programs were not beneficial to businesses, they would not be in existence today. The need for a well-defined program is essential, regardless of the size of the business concern or what is being protected.

Actively Communicate

The point to be made here is that there exists a necessity to educate the uninformed. If everyone possessed an awareness of the need for the time and expenditures made for fire protection measures, then these features would sell themselves. The more people who are enlightened on the subject, the better it is. What safety professionals know from studying destructive fires is that all fire loss is preventable. Actual failure to contain fire losses can almost always be traced to human carelessness, error, or oversight. The manager who flippantly ignores the need for ongoing fire safety falls smack dab into the "oversight" category. Due to a lingering residual of public indifference, the professionals involved in any facet of fire protection eventually realize that a part of every job they do incorporates sales.

To discuss the fundamental necessity for fire protection with a manager whose perspective on the subject is infinitely more shortsighted than your own requires an exchange that includes a planned presentation on your part. This presentation must be delivered in straightforward fashion, and outlined in advance. If nothing else, the amount of preparation made on your part will be obvious to the listener, demonstrating how seriously you feel that your views make good business sense. The correct approach then, is to formally set up a meeting for this purpose. Be persistent in this endeavor. If successful with this first step, remember that the power of a simple thank you is not to be underestimated, and just one phone call is all it takes to accomplish that gesture. In advance of the meeting, a genuine, "Thank you for agreeing to meet with me. I appreciate your taking the time. I look forward to discussing some of my fire safety concerns," and so on goes a long way toward the achievement of a rational two-way discussion.

According to sales trainers, 80% of the success in conveying the value of what you are discussing is determined by the quality of the presentation. A well thought-out, deliberately outlined, and organized proposal is what you're after. This is not gentle persuasion; there are important issues to be communicated. Have a plan developed. Expect resistance and prepare for it. Keep the presentation simple but substantial. Jot down the major points you wish to make on a piece of paper and bring them with you on a clipboard.

The dictionary defines "concept" as a general notion or idea. So remember that you're not selling widgets, you are *selling an idea*. With that in mind, you will have to allow for sufficient time for the "sale" to mature. Any steps taken toward increased fire safety are not going to be implemented overnight, so it's important that you plant the seeds of thought so that these steps may be budgeted and planned for the future. Because what you are discussing involves fire protection, maintenance, and engineering, a professional approach is recommended. Be punctual, polite, greet her with a smile, and then sit down and calmly discuss your objectives.[1]

Focus on the Positives

Success in any job is directly attributed to the attitude you possess. Be sure about what you are presenting because the degree of your own resolve will be evident. Also bear in mind that most people like to maintain a business relationship with someone who has a positive attitude. Restrict your remarks to those with a positive tone. Positives have a way of canceling out negatives, and positive attitudes are contagious. While not overstating things, your attitude should further convey an enthusiasm for how you feel your convictions and proposals will improve the workplace. Most important, the aspect of your attitude that should come across most of all at any meeting is your willingness to listen. As stated by James R. Fisher, Jr., "You have greater impact on others by the way you listen than by the way you talk." Listening is hard work, especially when you want to be thinking of what to say next, but don't forget that you have a clipboard at hand to refer to when the discussion of the other person's agenda has terminated. Anything less than your complete attention, and a willingness to stop for interruptions, will be looked at as a lack of concern on your part for the other person's prerogatives, opinions, and related concerns. Are her affairs important? You're darned right they are. And so are yours. This is how mutual respect is achieved, and it's the only way to make certain that your presentation works with maximum effectiveness.

Get Everyone on the Same Page

The next step is to involve key organizational members other than just a supervisor, director, or department manager. A successful fire protection program must incorporate the ideas of many key operations and business development personnel, taking into account their own goals and strategies. Although these efforts are time consuming, any good plan of action relies on teamwork, and giving due consideration to everyone concerned and affected by the action taken. The same process should also be put into place when interacting with the members of a community. It's been said that adults will remember 10% of what they hear and 60% of what they read. Because you have already identified a specific problem area, you can keep all colleagues current on the proposed plans and objectives through memos, e-mails, and other communications. Interoffice publications are another podium from which to be heard. Understand that any concept or proposal that people originally place low on their priority list must be followed up on as well, for that plan to prosper. And, because people are often hesitant to stick their own necks out, let them know that you are more than willing to take the responsibility and be accountable for implementation. It's not always easy to recommend change, but with the proper relationships significantly established, you can rely on the support you have created for yourself.

In-Place Fire Protection

If your place of business is not currently sprinklered, or is only partially sprinklered, properly installed automatic fire sprinkler protection for the premises is a top priority. Any fire statistics that you can put your hands on will always hammer down the known evidentiary fact that fire sprinkler systems are unquestionably the most effective avenue of fire protection for any structure. That major selling point notwithstanding, what I have always fallen back on is my own experience with system cost analysis. A small tool and die firm, with which I had business dealings, was once required by local officials to install a fire sprinkler system. Their total costs for this installation were high, in excess of $20,000 including the installation of a six-foot underground feed main. The fact that fire sprinklers save lives and property did not do much to diminish the owner's dismay over having been forced to borrow funds to pay for the retrofitted system. On my suggestion, he contacted his commercial insurance carrier to obtain a revised quote in lieu of the fact that he was about to sprinkler his property. He was surprised and delighted to learn that his annual premium was going to be lowered to $3300 from the previous total of $8100. In other words, his cash outlay for the fire sprinkler system would eventually be offset completely by insurance savings. He would be saving quite a bit of money beginning four to five years down the road. The point is that much higher fire insurance premiums are charged to businesses that are not protected by in-place fire protection systems.

Although the usual "payoff" time for sprinkler systems is more like seven to ten years for a building not previously sprinkler-equipped, it is a given that the lower building insurance costs allow systems to pay for themselves. Spread over a number of years, installing fire sprinklers becomes a self-liquidating transaction. And sprinkler systems maintain their value during resale of a building by making the facility more attractive to prospective buyers. One phone call to a commercial insurance broker is all it takes to sell this concept of fire protection, and your efforts will be applauded.

Ironically, businesses will install lawn sprinklers to nourish the landscaping, but initially resist spending money on automatic fire sprinkler protection that guarantees protection of the entire property and guards against business interruption. From a life safety standpoint, it's exhausting for many in the fire protection industry to even discuss the cost–benefit analyses, because they have seen that even one life lost is too many. We have had the technology for over 100 years to prevent the deaths resulting from fire, yet somehow many enterprises rely solely on the peace of mind bought with a fire insurance policy, rationalizing that that alone will safeguard both life and property.[2]

Fire sprinklers and fire loss programs are both integral parts of loss control. A total loss control effort, however, is not something that is purchased, but managed. With a greater understanding of loss potential comes greater cooperation and participation in the workplace. The security obtained in the form of protection from fire is also derived through the education and training of all employees. If nothing else, this education will be cause for employees to watch for hazardous conditions and be more conscious of their potential for danger. Whether the idea you are selling is for the purchase of fire protection equipment; the purchase of a fire protection system; the adoption of a new statute, policy, or ordinance; or the organization of a safety committee; you must be persistent, involve supervisors, promote and substantiate your ideas, and be satisfied with the knowledge that what you are doing will be of benefit to all. What you gain is security. It makes sense.

Endnotes

1. Mark Bromann, "Promoting and Selling a Concept," *PM Engineer*, June 2004, pp. 16, 20, 22.
2. Mark Bromann, "Promoting and Selling a Concept," *PM Engineer*, June 2004, pp. 16, 19.

4

The Local Municipality

What we in the private sector seldom do is sit down and have a conversation with someone who works at the Village Hall. Every now and then, fire protection engineers will have a brief discussion with the head of some local fire prevention bureau, but those are talks generally concerned with specific issues regarding requirements for a certain project. What the engineer doesn't hear about are the numerous problems the village administrator is dealing with herself, especially when the municipality in question has been or is currently caught up in a rush of growth. The problems are many for a city that is rapidly increasing in size.

False Alarms

Let's start with a common problem that has no clear solution. A small town often starts out with an alarm board at the police station, which will ring for any variety of alarms that are tied in to that central station. Initially, these will include burglar alarms, smoke alarms, carbon monoxide detectors, flow switches, and any variety of signals throughout the town that are tied in to the main board. As the town grows commercially in size, the magnitude of the number of local alarms in the town quickly make this station obsolete. The city may ask owners of fire sprinkler systems to tie alarms into a remote station outside of town, as a quick fix to alleviate the problem. But regardless of logistics or method, false alarms will be a guaranteed expensive nuisance. There are fewer (commercial) fires today, but more false alarms.

A large Chicago suburb near O'Hare Airport receives about 1,100 false alarms yearly. The causes of these, according to that town's fire chief, are roughly 60% system malfunction and 30% "stupidity." Denver, Colorado reports that they receive well over 10,000 false alarm calls a year. The problems associated with this aren't only about cost. From 1984 through 1993, statistics document that 26 U.S. firefighters were killed while responding to false calls (11 of those attributed to alarm system malfunction). There is no easy solution to this problem. The whole idea of a fire alarm system is for the city fire department to respond as fast as possible to a fire. There is no such thing as an alarm panel that tells you which alarm is false, and personnel may be unable to respond effectively to a real fire if out on a false alarm.

The National Institute of Standards has reported that there are 16 false alarms for every real fire. A survey of fire chiefs yielded the estimate that there is one false alarm per year for every 100 detectors in use. Of the estimated 1.6 million false alarms in the United States, the causal factor breakdown goes as follows:

Malfunctioning fire alarm systems   41%

Malicious (mischievous) calls   25%

Unintentional (accidental) calls   24%

All other, including bomb scares   10%

As anyone who has had a smoke detector go off while showering will attest, each nuisance alarm make it more likely that people will ignore future alarms. Worse, an accumulation of false alarms often induces people to remove the batteries from their detectors. As far as Merton Bunker of the NFPA (National Fire Protection Association) is concerned, "Between better installation and maintenance, many nuisance alarms could be dispensed with." When cities receive multiple false alarms from one building, many turn to fining the offender. Of course, this action may also lead to building owners disconnecting their systems altogether. But it is hoped, as fines accrue, building personnel will direct that flow of funds toward alarm system repair or upgrade.

Some villages tackle the problem by amending their own city code to try to come up with solutions that are effective. For example, some require heat detectors as opposed to smoke detectors simply because they malfunction less frequently. Other towns have dropped the requirements for manual pull stations in stores that are sprinklered, or (not recommended) have dropped the requirement for detectors completely if the property has sprinklers installed throughout.

Other municipalities seemingly do nothing about the problem, and unfortunately the motive there is often of a financial nature. Consider the existence of city funding: every city department seeks an appropriate piece of the budget pie. And naturally, the larger number of responses serviced by a "paid" fire department, the more funding they receive. It is representative of more work. This is why the fire department usually responds to a "wires down" call. They want those calls: this documents the added work the typical firefighter is doing, and each additional firefighter needed on the department represents a bona fide funding increase.

But speaking frankly, most departments want fewer responses and despise false alarms out of general principle. In Toronto, almost half of all their false alarms once came from one inner-city ward consumed with gang activities. The city eventually had to remove all the pull stations from the public corridors of those buildings. Philadelphia has done away with their "street boxes," due to numerous false calls and the existence of the improved 911 system. These changes are expensive, but they are common-sense solutions that have worked.

Legislation

Many public officials often complain about the state legislation that they must enforce. Typically, these state mandates have to do with accessibility issues such as aisle and door widths, or with backflow prevention, and building reviews. Although the local personnel may not always agree on a requirement for say, a public restroom in a 200-square-foot train depot, the state code (someone else's rules) must be enforced. And the state doesn't compensate the city for enforcing these rules.

The legal key to fully understand the actions of municipalities is their own enacting legislation. Does it give them the authority to enforce or interpret the model codes? About 90% of the time, it's just to enforce the codes, but you have to carefully read the wording of any local legislation. A power game is always going on. The NFPA has always maintained political acceptance of AHJs (Authority Housing Jurisdictions), in part so that they will adopt applicable NFPA codes. Most cities adopt building codes (specifying this by yearly edition) to avoid complication. But complexity is added when the city writes and adopts its own code amendments. These often represent instances of the town going a little overboard, and there are many examples of this of which we are all well aware. A Chicago suburb has recently mandated sprinklers above drop-ceilings for noncombustible commercial construction. Not that this will hurt anything, but it is obviously unneeded and represents an expense that produces no real benefit. How about this suggestion: since the big problem is residential fires, and about 42% of apartment fires originate in kitchens, and most of those are started by a cooking fire that ignites the wooden cabinet above the stove, why not address that reality? A city code amendment could give the builder the option of either (1) installing one sprinkler near the kitchen stove or (2) not installing any cabinet directly above the stove. You know, fight the real battle.

Plan Review

The municipality has to strive to avoid getting the reputation that is hard to work in that town. Once they have such a reputation, they only get the lousy contractors and this is a bigger problem that is hard to alleviate. By following some simple guidelines (and avoiding pitfalls) with regard to new construction, the municipality can establish and maintain a healthy professional standing:

- Involve the architect; this is her project.
- Let her know right away what codes have been adopted, and of any unusual local requirements.
- Review what is submitted, as quickly as possible. Avoid the rejection of plans without a conscientious review.
- The degree of involvement should be synonymous with the magnitude of the project.
- Let the GC manage his own job, so he cannot blame the city for delays. GC problems are GC problems, and not necessarily City problems.
- Return calls quickly.
- Reprimand any city personnel with an antagonistic attitude, and remember that everyone appreciates common sense.
- The building and fire departments must communicate and all departments must work together without dissension.
- Establish a deadline for variance applications.
- Be aware of any plans for future expansion.
- Make sure that submitted plans are up to date.
- The plan review should not contain too many "show me" comments (i.e., indicate this, label this, explain this, etc.).
- Focus #1: The hydraulic calculations for fire sprinklers must work!
- Focus #2: The final inspection must be thorough and conclude that the installation is an accurate representation of what has been indicated on the plans.[1]

Existing Construction

Older towns contain sections or blocks of properties built at a time when compliance with building codes was of low priority. Although these areas are rich in charm and appeal, they constitute a nagging worry to fire chiefs. Sprinklers may be nonexistent, and the construction is high in combustible materials. Fortunately, a fire station is normally close by, but what is of great concern is the potential for widespread fire. Of equal concern is the fact that there are often apartments in floors above these downtown businesses. Retrofit sprinkler ordinances usually meet with such fierce opposition that efforts to try to pass these are almost always fruitless. Research has indicated that fire "flashover" can and will occur in older typical wood frame structures.

What the fire departments desperately need in these situations is early warning. Villages normally ask for heat detectors throughout these premises. One other effective action that can be taken is preventive in nature: that of an aggressive fire safety inspection program. What cities do in this case is first identify the boundaries of the business district area in question. An inspection report form can be used, but what is more important than its format is its content. Items to be covered in the inspection report include:

- Information on emergency notification
- Minimum storage clearances
- Fire doors and exit facilities
- Knox box installation (if required)
- Clearance around heat sources
- Condition of flue/vent piping, chimneys
- Flammable liquids storage
- Existence of any open wiring
- Accessibility of utility shutoffs

This is a short list, of course, and a much more comprehensive inspection is needed from either an on-duty member of the local fire department, or an outside consultant. An inspection may turn up any number of violations from frayed extension cords or overloaded electrical outlets to out-of-service emergency lighting. In general, inspections cover fire safety in the operation of the building and its occupancies, maintenance of fire protection devices, and maintenance of mechanical and electrical systems. An effective inspection campaign will also increase public awareness and, it is hoped, foster fire education programs. See Figure 4.1.

Pre-Fire Plans

As a part of each fire inspection, an existing pre-fire plan can be updated, or a new one prepared. This increases the likelihood that the inspector will make a complete tour around the building. After that, someone computer-knowledgeable will normally use a standard format to develop a plan for each property that can be referenced inside the fire engine as the department responds to a call. The stored information is organized in an alpha/numeric system based on property address, with a goal of plan readability as well as accuracy.

The formatted plans show a scaled representation of each property. Standard symbols on pre-fire plans delineate various key locations such as

FIGURE 4.1
Per local requirement, the outlets to this fire department connection have been equipped with locking Knox caps.

post indicator valves, sprinklered areas, standpipes, fire department connections, sprinkler risers, detectors, fire doors, roof hatches, fire pumps, sprinkler valves, Knox boxes, skylights, loading docks, man doors, fences, utility shut-offs, wall hydrants, annunciator panels, and so forth. This means that familiarization of property and structure by the fire department is an instrumental resource for municipal fire protection. Organization and quality are paramount ingredients for pre-fire plan success.

Outside Consultants

When a city experiences expansion, this growth is best managed in-house, and more inside personnel must be hired. But growth is difficult to gauge, and an orderly balancing of time and responsibilities is a tough task when labor demands cannot be accurately forecast. Outside consulting fees are not as astronomical as they seem when you consider that the consultant does not use a village automobile, consume office overhead, or pose a cost in terms of payroll taxes or insurance. Nor does he receive vacation, sick leave, or holiday pay.

Villages often seek to lessen burdens on city staff members, and in the process save base labor costs through the employment of outside assistance. Consultants are routinely hired to perform plan reviews, inspections, and also to prepare pre-fire plans. Communication is essential, and the municipality strives to find people who are able to work in conjunction with the instructions and needs of the village staff. The use of consultants is particularly advantageous during periods of high activity. Also, in the spirit of the "three heads are better than two" ideology, city personnel usually acquire additional information and methodologies from the consultants that they would otherwise not have learned.

The biggest complaint that I have heard from municipal officials about consultants is when their work is not up to par. Outside consultants must be competent, licensed, and experienced. No one just becomes a consultant: actual field experience is a plus. The best plan reviewer, for example, is an individual who has actually designed plans himself. Surely, an inspector should have documented well-rounded experience in conducting inspections. And, the consultant must recognize that he is the consultant and not the AHJ.

Regarding building plan review, it is much easier to hire one consultant to review all plans, but not always wise. Using different consultants to review the work of different trades, especially fire protection, is a much more prudent approach. Cost is another factor. Some agencies such as the ISO (International Standardization Organization) or BOCA (Building Officials and Code Administrators) may offer more reasonable fees than a single plan review agency. Sometimes a "friend" who happens to be a contractor proficient in one area may offer to help out the city in reviewing plans or calculations. Regardless of cost, this scenario is to be avoided due to obvious conflict of interest concerns. Certain municipalities will require plans to be stamped or approved by a P.E. or someone NICET (National Institute for Certification in Engineering) -certified, which is another way to safeguard legitimacy. In any case, what needs to be ultimately investigated is the cost of municipal fire prevention, and this can only be accomplished through an in-house, in-depth study.

How the fire service responds to the needs of its constituents speaks volumes about its integrity. Regarding false alarms, everyone should desire fewer calls and take steps toward achieving that goal. Code adoption is vital, and decisions in this arena are not to be rushed. Specific needs particular to any one community can be addressed through the scripting of local codes. The city water supply must be absolutely reliable and well-maintained. Plan reviews should be professional, focus on major issues, and have a quick turnaround time. Although the quaint beauty of the older sections of town is to be treasured, efforts toward fire prevention for those districts should be doubled. Time and planning in terms of fire prevention in the form of inspections, pre-fire plans, and public education are absolute necessities. Outside consultants should be contracted for professional services if they are dependable and competent, although huge consulting firms who handle many communities may not always be the best answer. Within those firms, certain consultants are much better than others.

Fire departments exist in a real world where a variety of purposes is to be addressed and served. Especially in consideration of future municipal growth, local administrators have increasingly addressed the needs of multiple functions through the concept of master planning. This is a fancy way of stating their goal of seeking the most cost-effective allocations of resources to ensure that the needs of all community elements will be met. After the fire protection problems of the jurisdiction are identified, as well as its current capabilities, an all-inclusive plan is devised that includes goals and objectives. What follows are planning, implementation, and then feedback. It's a never-ending process that must be continually updated.[2]

Endnotes

1. Mark Bromann, "Problems Facing Municipalities," *PM Engineer*, May 2000, pp. 28, 30.
2. Mark Bromann, "Problems Facing Municipalities," *PM Engineer*, June 2000, pp. 24, 26, 28.

5

The Role of Firefighters

In theory, the objective of an automatic sprinkler system is to control an outbreak of fire until such time as the firefighters and their apparatus arrive. In an urban or suburban setting the timeframe for that arrival is normally between four and six minutes after notification is received. The rural area population does not enjoy the same quick response time. However, history has recorded that Americans living in urban or rural areas have a much higher risk of dying in a fire than those living in suburban areas and small towns. Each year, fire kills more Americans than all the major natural emergencies combined, including floods, hurricanes, tornadoes, and earthquakes.

How much worse would the toll be were it not for the fire service industry, one of the familiar securities that many of us take for granted. Whether they be volunteers or members of a full-time paid department, our nation's firefighters operate as a team. Typically they serve us on a 24-hours-on, 48-hours-off basis. Although on the average they will battle less than a dozen fires annually, they cannot afford to sit back and wait for a fire to occur. Their duties are far more complex.

Firemen normally train at least two hours per day in firefighting technique and organization. They coexist with a fire prevention bureau, a staff of individuals who work a more customary 9-to-5 job. Together, their time is spent pooling their resources into prevention programs and inspections, so that responding firefighters have an edge when called to public service. One of the chief objectives is to ensure that buildings are compartmentalized with fire walls, firestopping or parapet walls, and fire-rated enclosures. Streamlining construction in this manner ensures that there will be smaller areas in which to fight fire. Protective clothing worn by responding firefighters today is extensive, including a helmet, boots, gloves, coat, and self-contained breathing apparatus (SCBA). They carry hose harnesses, axes, and often a "Denver tool" to pry apart portions of structure to locate smoldering fuel. See Figure 5.1.

The fire prevention bureau endeavors to work with architects and developers in the planning stages of building design to validate that safety features are installed in new construction. If the efforts and resources of fire department personnel are successful, then local structures will have such amenities as early detection and alarm systems, automatic fire sprinklers, standpipes, hose stations, column sprinklers, and the like, so that the firefighter, when summoned, can maintain a constant advantage over a potential inferno. Local fire prevention bureau personnel schedule periodic building

FIGURE 5.1
This 1998 HME/Luverne engine includes a 1,500-gpm pump, with a 500-gallon tank (photo by Mike Charnota).

inspections, which are conducted during regular business hours. It is advisable that building owners maintain an unblemished relationship with their fire department at all times. Landlords, business owners, and fire system designers must work hand in hand with fire officials.

In the fire station, there is an inherent need to build teamwork. The individuals that make up the fire companies develop their own camaraderie in two ways. First and foremost, they are a family always looking out for one another in a chain of command, or buddy system, when they must fight dangerous fires inside weakening structures. The camaraderie is further enhanced by spending down-time in the fire house or elsewhere.

There is an establishment in Itasca, Illinois called Ben's Tavern that still has a fire alarm behind the bar. That bell, once wired to the fire department, is a holdover from the days when drinking was a more socially acceptable norm during the workweek. The bell's function was to alert firefighters—playing cards with mugs in hand—of an emergency situation. Today modern firefighters spends much of their down-time involved in activities such as continuing education, training drills, and physical exercise. All of this work serves to enhance team readiness. In the face of fire, a drop of water can extinguish a match flame. A gallon of water can put out a wastebasket fire. But when fire gains the momentum of an inferno, an entire river of water cannot stop the blaze. So, the importance of the team energy cannot be minimized. See Figure 5.2. Surely there

FIGURE 5.2
This 1981 Ford/Emergency One engine includes a 1,250-gpm pump, with a 500-gallon tank (photo by Mike Charnota).

are lives at stake. No form or amount of insurance can replace the consequences of catastrophe.

The first known organization of American firefighters put its noble bunch together in Boston in 1679. No one can fight a fire alone. This has always been a group effort and is part of Americana. Fire confinement and extinguishment are the duty and responsibility of the fire department. The first person alerted to a fire (or smoke) in a building must prioritize life safety. His or her responsibility is to leave the fire area, shut the door, and take other occupants along as 911 is called. Then, proceed to the nearest stairwell or escape route, pull a manual alarm, and exit the premises. If a large amount of black smoke has been generated, people should lay low. Noxious fumes produced by fire in modern buildings can be extremely toxic and can affect the central nervous system or the cardiovascular system in anyone.

There is a chain of command implemented for emergencies, and the probability of that success can be anticipated by the overall measure of preparation. Comprehensive education and training is conducted frequently within the corps. Fire brigade members possess knowledge of fire suppression equipment, management support, a skilled understanding of fire behavior, knowledge of ignitable products, and knowledge of building design, and they maintain ready equipment. They are safety experts, paramedics, safety planners, and loss prevention educators. They serve the public and ensure the continuity of industry. Fire safety is not just about emergencies, but continually developed as a sound program of prevention.

A fire department may exercise its authority when situations arise that constitute potential disasters. A place of public assembly, for example, may expect a large attendance at a concert or special event. Promoters may decide to actually lock exits, attempting to prevent a ticketholder from opening a remote door to allow entry by a nonpaying customer. Because this action amounts to nothing more than locking people in a building, an abundance of firefighters placed in and around the structure during the show or concert will serve to enhance public safety. Similarly, there exists the seasonal temporary structure known as the Haunted House. These are dangerous places. Randomly plunged into a vacant home are: temporarily converted open-wire sound systems, with hastily constructed particleboard walls and corridors, complete with a plethora of cramped combustible load, all in a dark structure hoping for a maximum capacity of human occupants. Haunted Houses contain just about all of the ingredients for a potential catastrophe. Electrical codes have not been followed. Exits are hard to find. Paint cans are lying about. There are workers operating a sound system in the attic. It all adds up to a horror story for those in fire prevention, so once again a firewatch is organized, which is why you will often see many firefighters stationed around such an occupancy in late October.

A sobering fact is that a small percentage of rescuers in all types of emergency situations are fatally injured as a direct result of their own rescue efforts. In the United States, roughly 1 out of 65 lives that are lost in a fire are those of firefighters. That's a little over 100 per year that perish in the line of duty. We may take them for granted, but our dependence on them is complete. A menace we must live with has always been, and always will be, the potential for fire.

More than two million fires are reported each year in this country, and millions more go unreported. Direct property loss due to fire is astronomical. That's the overall picture, and it goes on year after year. The firefighter is an educated professional, an expert in the knowledge of where fires are most likely to occur, what people are most at risk, and how to save lives. Their heroic service is not unnoticed.[1]

Endnote

1. Mark Bromann, "A Tribute to Firefighters," *PM Engineer*, December 1998, pp. 22, 24, 26, 28.

6

Automatic Sprinkler Systems

An automatic fire sprinkler system has been defined as an integrated system of overhead piping, designed in accordance with fire protection codes and standards, to which automatic sprinklers are attached. The fire sprinklers contained within the system will be activated by the heat from a fire, which will immediately discharge water over the area in which the heat was received. Water is supplied to the system from tanks or underground water supplies. If the building being protected is situated in a low (municipal) water pressure zone, or if additional water pressure is required due to the overly hazardous nature of the building's contents, then the presence of a fire pump will become a necessity within the structure. A fire pump often becomes critical in order to meet insurance carrier criteria or the standards of the national fire code. Sprinkler system water supply requirements are based on the type of building construction along with the degree of hazard presented by the specific building occupancy. If a fire pump has been installed, the sprinkler system's water supply will receive a pressure boost from the pump when system pressure begins to drop during a fire. The objective of all sprinkler systems is very simple: to keep people and property secure and protected from the threat of fire when fire occurs.

Should the sprinkler system cover less than 5,000 sq ft, system piping may be sized per National Fire Protection Association (NFPA) standards to conform to a pipe schedule that assigns a particular pipe size based on the number of sprinklers supplied by the pipe, which will vary based on the degree of hazard of the occupancy. More often, the pipe sizes are hydraulically designed, tailored to the specific occupancy being protected. There is a maximum square-foot area that each sprinkler system may protect within a structure, so for buildings covering a large expanse, there may be more than one sprinkler riser installed within. Each riser contains a controlling valve, pressure gauge, a water flow–indicating device for alarm purposes, and a main drain valve. The systems are normally designed so that all water from the system will drain back to the main drain valve, typically through a 2-in. angle valve that will discharge drainage water to the outside.

Wet-Pipe Systems

In the most common sprinkler system type, the wet-pipe system, all piping contains water under pressure. This is the most reliable system type and is recommended whenever practical and wherever the system will be installed in a climate-controlled building. Provided that the supply valve and the riser valve are in open positions, all that needs to operate for immediate fire extinguishment is the sprinkler itself. To debunk one persistent myth, we need to understand that only the sprinkler or sprinklers above the fire (which have fused) will operate during a fire. When they do, a device (usually a flow switch) will activate an alarm that will sound on the premises and also automatically notify the local fire department of the situation. The building owner is responsible for the proper maintenance and prime condition of the sprinkler system he owns. As with all systems, some type of check valve or backflow prevention device must be in place to prevent the stagnant contaminated water within the sprinkler system from somehow flowing back into a city water supply. The system also contains a remote inspector's test valve that is used to test the flow switch. An orifice placed in the inspector's test discharge piping mimics the flow of one system sprinkler.

The major advantage of a properly installed sprinkler system is that, because sprinklers are strategically placed at intervals along the piping system, the discharged water can always reach the seat of the fire. At the same time, rising smoke is mitigated by the downward force of the water, which gives room occupants ample time to remove themselves from the point of initial disaster. It should be noted that a sprinkler system that operates in a timely manner discharges much less water than would be generated by the hose stream discharge from fire department hoses. The system operates only in cases of intense heat, only in the fire area, and cases of accidental release of water from sprinkler systems are extremely rare. See Figures 6.1 to 6.3.

It is preferable in almost all cases to have the inspector's test connection placed at a remote section of the system. This valve is kept open during system refilling operations, and should be slowly closed as water discharges. This exercise ensures that a sufficient amount of air escapes from the system before it is placed back into service. Too much air that stays in the system during repeated inspections and drain-down operations will hasten the development of corrosion within wet-pipe system piping.

Regardless of system type, all system control valves are electrically supervised through the use of tamper switches. These valve position indicators will activate a trouble signal should the control valve be closed. Inasmuch as closed valves represent 30% of all sprinkler system failures, it has become

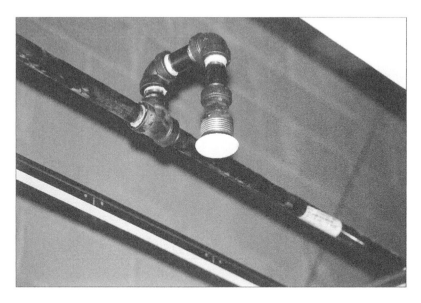

FIGURE 6.1
This "concealer" type pendent sprinkler has been installed on a return bend for a center-of-tile installation.

FIGURE 6.2
In instances where only a non-combustible concealed space will exist above a drop-ceiling, automatic fire sprinklers need only be installed beneath the new ceiling, and not above it.

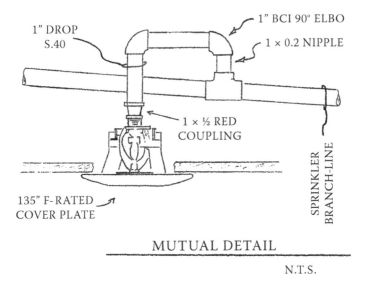

1" DROP
S.40

1" BCI 90° ELBO

1 × 0.2 NIPPLE

1 × ½ RED
COUPLING

135" F-RATED
COVER PLATE

SPRINKLER
BRANCH-LINE

MUTUAL DETAIL

N.T.S.

FIGURE 6.3
A configured detail of a return-bend sprinkler installation.

necessary to mandate valves of the indicating type for all control valves located within a sprinkler system, and to electronically monitor their open position or at the very least, require that the valves be locked open by mechanical means (see Figure 6.4).

All sprinkler systems consisting of 20 or more sprinklers will be equipped with a fire department connection that is mounted on the outside of the building, preferably within 100 ft of a fire hydrant. The fire department connection piping, usually 4-in., contains a check valve to prevent water from discharging from the connection itself. A bell is mounted outside, above the fire department connection, to help firefighters locate the connection when necessary. Sometimes this is an audible/visible strobe device, which is especially useful in instances where the fire department connection may be covered by snow or if its view is obstructed by shrubbery. Code requires that the fire department connection be placed at an elevation between 18 and 48 in. above grade. It's important to remember that in the event of an accidental valve closure, the fire department connection then becomes the sole water supply for an automatic sprinkler system. See Figure 6.5.

FIGURE 6.4

This wet-pipe system riser includes two O, S, and Y valves with tamper switches, a double-check detector assembly, pressure gauge, flow switch, 2-in. main drain, and a spare-head cabinet mounted nearby. (See color insert following p. 52.)

FIGURE 6.5
A single-outlet fire department connection, permissible for small systems that are 2-1/2 in. or less in size.

Dry-Pipe Systems

By appearance, a dry-pipe system looks identical to the wet-pipe system, but all dry system piping contains air under pressure that is maintained by an air compressor. These systems are employed where the building may be subject to freezing temperatures. Upon sprinkler activation, the system loses air pressure quickly, and the subsequent loss in air pressure (below a predetermined point) trips the dry valve (at the riser) which contains a clapper that has been previously held shut by the air pressure. The system then fills with water that soon reaches the opened sprinkler(s). The resultant fire extinguishment will be accomplished, although the dry-pipe system will be somewhat slower in terms of overall reaction time than a typical wet-pipe system. For this reason, if part of a building is climate-controlled and another area is unheated, it is recommended that the dry system protect the

FIGURE 6.6
A close-up of a dry system riser showing the dry pipe valve and trim. (See color insert following p. 52.)

cold area and a wet-pipe system be employed for the rest of the structure. The intended use of sprinkler systems is to deliver water to a fire as soon as possible. For a dry-pipe system to be acceptable to authorities, the system must be able to discharge water from its inspector's test connection within 60 seconds of the opening of that test valve.

The dry valve itself is a latching differential valve that is situated in the system riser immediately downstream of that system's control valve. See Figure 6.6. It is always installed in the vertical position and must be installed in a heated room. The valve may be installed in conjunction with an accelerator or other quick-opening device used to increase the operating speed of the dry valve assembly. This component will be installed in the trim of the dry valve, typically on half-inch galvanized steel piping. A flow switch is not used on dry-pipe systems. Instead, a pressure-operated electric switch will operate when the dry valve is engaged, thus constituting an alarm appliance that initiates directly from the intermediate chamber of the dry valve. This pressure switch is supplied by the manufacturer of the dry valve for the purpose of electric water flow alarm operation.

The compressed air supply is normally taken from an air compressor installed near the dry valve, also in an area capable of permanently providing sufficient heat. Proper air pressure must be maintained so that the compressor is not kicking on more than once weekly. The compressor is an electric motor-driven, air-cooled component. System air pressure shall be maintained in accordance with the manufacturer's data sheets, or roughly 20 psi in excess of the calculated trip pressure of the dry valve. If a building contains just one system and that system is dry-pipe, the 4-in. feed from the fire department connection should connect in between the system control valve and the dry pipe valve. The main system drain is piped from the lower portion of the dry valve. The drain valve is sized to 2 in. for all system risers that are 4 in. or larger.

Installing a dry-pipe system will incur additional material and labor costs to provide the dry valve and trim, compressor, and electrical wiring. As previously mentioned, the primary disadvantages of the dry system are the longer lag time for water to hit a potential fire after a sprinkler fuses, and the fact that there are more chances for some type of mechanical failure on a dry as opposed to a wet system. In addition, any pendent sprinklers used on a dry-system are more expensive to purchase. These "dry pendent" sprinklers are used if installed in a nonheated area and are made up of an elongated piece of "vacuum" piping that attaches to the sprinkler itself. The dry pendent sprinkler extends upwards into the sprinkler system branch line (to avoid "ponding" of the water in the fitting above the dry pendent). All system piping on a dry system must be properly pitched according to NFPA code requirements so that the system can always be effectively drained. Any "trapped" piping (in system low points) must be accompanied with system drain assemblies, which must have visible signage attached.

Deluge and Preaction Systems

All sprinklers are "open" in a deluge system: they have no glass bulbs or fusible elements. The piping is dry, containing air that is not pressurized. With a deluge system, fire detection units are installed alongside the deluge system piping and, when activated, the deluge valve will open and water then discharges from all open sprinklers, or "open nozzles." The detection or "release" system serves all areas that the system protects. If the system is preaction, the piping is still dry but with pressurized air, and only standard conventional sprinklers are used. Thus with the preaction system, detection system activation serves to fill the system with water that will only discharge if and when one or more sprinklers operates. The choice to install either system depends on the characteristics of the occupancy being protected, and not necessarily those without adequate heat.

In either case, a deluge valve prevents water from entering the system piping until such time as detection is activated and the valve trips. There is a pressurized upper valve chamber on a deluge valve that keeps the valve closed. It is also connected to the release line, which will relieve pressure from the upper chamber upon activation, thereby allowing water pressure to lift the clapper inside the deluge valve and flood the system. This operation also rings an alarm.

The deluge system is most commonly employed when it is advantageous to spray water from all nozzles to protect a clear-cut high hazard, such as an industrial press, or an inside or outdoor system of piping, tanks, or vessels containing highly flammable materials. The preaction system is most commonly employed in highly decorative or highly sensitive areas such as museums or computer rooms, in which the application of water is not desirable and must only be discharged when there is little chance that the condition experienced is something other than a highly intense fire that must be controlled. Because the electrical (or pneumatic) actuating release system will activate prior to the operation of sprinklers, water application is actually accomplished more quickly with a preaction system than with a dry system. But this is still a special hazard application and carries with it additional system components that must be periodically inspected and tested. No more than 1,000 automatic sprinklers may be controlled by one preaction valve or automatic water control valve. The alarm devices used for both deluge and preaction systems are similar to, if not the same as, the ones used for dry-pipe systems. Installation guidelines for both preaction and deluge systems can be referenced in Sections 2.4.4 and 2.4.5 in *Loss Prevention Data Sheet 2.0*, published by FM Global.

Water Mist Systems

Water mist systems consist of a network of stainless steel piping that supplies water to nozzles which produce tiny atomized droplets to totally flood an enclosure. By blanketing fine-spray water droplets over a wide surface area, an activated water mist system will maximize heat absorption and displace oxygen. A major benefit of this type of system is rapid cooling, which is essential for fire mitigation (and the "blocking" of radiant heat) in areas where machinery cannot quickly be shutdown. Water mist systems are installed in large computer server rooms, power generation utilities, historic structures, underground railway systems, oil and fuel pump rooms, electrical equipment rooms, and any other occupancy where an abundant amount of water is not desirable. They typically employ high pressure (and low volume) pumps to deliver water. The relatively small (1/2-in. to 1-1/2-in.) stainless steel piping is designed to hold pressures exceeding 1,000 psi. As an option, copper tubing may be used in lieu of stainless steel. Water mist systems require a pristine water supply, ideally potable water. A water supply that consists of a tank that is filled with well water is unsuitable.

The fire detection system used consists of 3/4-in. CPVC piping containing small holes that continually draw air by means of a high-efficiency aspirator, passing a sample through a filter before it enters a laser detection chamber for smoke detection. Although much less water is discharged in a fire scenario, total flooding will occur because the droplets are drawn to the base of a fire. As the water mist turns into steam (expanding in volume), it depletes the oxygen necessary for continued combustion. This formula for extinguishment also prevents reignition. These systems efficiently absorb and remove both smoke and toxic gases from a fire, a feature not present with alternative system protection options. In most cases, water mist systems are prohibited in areas with a ceiling height that exceeds 15 ft, or that will have an enclosure volume in excess of 60,000 cubic ft.

Water mist systems are most effective at extinguishing fires involving common electrical materials. Of equal importance, testing has demonstrated that water mist will not harm computers even after the system has discharged for 30 minutes. Water mist systems are a very expensive fire suppression option, however; these systems will disperse only 30–50% of the water that a sprinkler system would deliver to suppress the same type of fire.

7

Water Supplies

The BOCA (Building Officials and Code Administrators) National Fire Prevention Code defines an automatic water supply as water supplied through a gravity or pressure tank or automatically operated fire pumps, or from a direct connection to an approved municipal water main. Requirements for water supply capacity and water supply duration are found in Chapter 11 in the 2010 edition of NFPA (National Fire Protection Association) #13, *Standard for the Installation of Sprinkler Systems*. The available water supply must be a known factor so that the fire suppression system may be tailored and suitably designed to function reliably with that supply. A direct piped connection to a municipal water supply is normally used when one of sufficient volume and pressure is available. If a municipal water supply is unavailable, or too far away from the building, stored water in the form of ground storage tanks, reservoirs, lakes, rivers, gravity tanks, or some other type of tank will be implemented to provide the expected sprinkler system demand for whatever duration of time is dictated by code. In these instances, a fire pump will be necessary unless the stored water can be gravity fed.

For a typical ordinary hazard occupancy (metal working plant, laundry, bakery, parking garage, repair garage, retail establishment, etc.), regardless of the type of automatic water supply, the NFPA standard requires a 60-minute duration of water to supply the hydraulically designed demands of the automatic fire sprinkler system, plus 250 gpm for outside hoses used by the responding fire department. Most public water supply systems are designed with a dual purpose in mind: that is, to supply water for (domestic usage) drinking, sanitary, and industrial uses, and also to provide emergency water for fire hydrants and to supply fixed fire suppression systems (see Figure 7.1). Although the adequacy of a public water system cannot be taken for granted, it was noted years ago that in large cities which required a substantial volume of water for domestic purposes, there was usually adequate water provided in the same existing water supply to provide for fire emergencies. A fire pump may be required to boost city pressure to meet system demands, particularly for taller buildings.

Most automatic fire sprinkler systems utilize a municipal water system as their only source of water supply. Obviously, there needs to be an established quantity of water for fire protection use. Any fire suppression system will be designed to a known water pressure figure with an accompanying flow provided by the water supply. In order for the fire sprinkler system designer to complete the exercise known as the hydraulic calculation, the pressure and

FIGURE 7.1
Municipal workers put the finishing touches on a fire hydrant installation.

volume characteristics will have had to be determined for the water supply. Only in this manner can the volume and pressure demands of the particular sprinkler system be ascertained as adequate and approved, and proven to be within the scope of what the local water system can supply. Of course, to determine the characteristics of a particular area of a municipal waterworks system, a flow test must be made on that system.

Water Flow Testing

It is the rare exception when a designer ever calculates a fire sprinkler system without first being cognizant of the local flow test results, and how the resulting plotted "curve" looks when drawn on N1.85 hydraulic graph paper. By anticipating the total system demand, the designer can easily guesstimate what the available pressure will be, and then simply tailor the hydraulic design in such a manner that will leave a 3 to 5 psi "cushion" between the final system demand point, and what volume of water (in gpm) at what pressure (in psig) is known to be existent at that precise building location.

Suffice it to say, the flow test is usually made with all intentions of being as accurate as possible, even though it is at best an approximation, taking into consideration the various factors such as the integrity of the equipment used, the time of day at which the flow test was taken, and human error. Engineers sometimes forget that when human beings are involved, nothing is going to be always absolutely perfect. But, by virtue of design, hydrants are capable of discharging a measurable quantity of water at a measurable pressure. We must also consider the time of year that the flow test is taken. To address a worst-case scenario, the flow test should probably be taken during either July or August, which are the most demanding months. And how about time of day? Certainly, a 6:00 p.m. test will yield results that are more indicative of the worst-case scenario than will a test taken at 11:00 a.m.

But let's take a look at what the sprinkler system designer may be given to work with. The local water purveyor has faxed over to him the following information summarizing his recently conducted flow test: a static psi read of 54 psi, and a residual pressure reading of 36 psi at a total flow of 839 gpm. The general rule-of-thumb says that as long as there is a minimum pressure drop of 10 psi, the flow test results are accurate enough, and there is no need to open more than one port, or the steamer, on a fire hydrant or more than one fire hydrant. In this case, the employees of the village in question probably opened one 2-1/2-in. port full bore, and got a reading of 25 on the pitot gauge. I say this because the formula for total flow is: (29.83) [coefficient] (size of outlet squared) [X], where X = pitot reading. So the municipality in question simply multiplied the constant (29.83) times the .90 coefficient, times the square of the nominal 2-1/2-in. outlet (6.25), times the square root of the pitot reading (5), to reach the total flow figure of 839 gpm. The majority of fire hydrants in the greater Midwest have a smooth, rounded, discharging outlet (which correlates to a .90 coefficient). This factor must be taken into consideration when deciding which coefficient is to be used when determining total flow in the aforementioned calculation.

The results of this flow test are handily plotted on the N1.85 graph paper. The horizontal coordinates of the graph sheet are scaled to the 1.85 power because, in the Hazen–Williams formula, pressure in psi is proportionate to the flow in gpm to the 1.85 power. In order for the designer to draw this curve most accurately, the city has supplied a third "point" for this curve, by stating that at a residual pressure of 20 psi, the total flow will be 1,182 gpm. This information represents not actual test results, but a reliable approximation that they have obtained through the use of a computer program. So, with the click of an index finger on a mouse, the computer has come up with a third point to plot on the water supply curve.

Just because a building is protected by an automatic fire suppression system, it does not necessarily follow that there is a carte blanche guarantee of the effectiveness of the sprinkler system. The system must be supported by a water supply that can meet the demands of both the flowing sprinklers and the accompanying hose streams during an actual fire. Earlier, I mentioned the designer's proclivity toward the hydraulic design of the sprinkler piping to within 3–5 psi of the available pressure noted by the supply curve. There are several reasons for his doing this that makes it a good common-sense practice. For one thing, the system may not be installed exactly as it is designed. The shifting of pipe here and there, adding a few offsets, and perhaps the necessity for riser nipples when none were originally designed, will adversely affect the outcome of the actual water pressures when delivered. Second, the water flow test used may be out of date. It is always only a close approximation, but what we do know is that many water supply systems deteriorate with age. (That rule also applies to the fire sprinkler system piping itself.) Third, we must consider the design life of the building being protected, which is typically 75 to 100 years. Over that span of time, occupancy changes are likely to occur, and there is no guarantee that if some tenant change occurs 40 years down the road, that commensurate modifications will then be made to the sprinkler system. So, although there is nothing noted in NFPA standards regarding such a buffer, it is wise for the hydraulic designer to recognize a 5 psi margin of safety consideration. I should add that this cushion will represent a much more conservative approach if the water supply curve is nearly flat, but can mean very little if the slope of the curve is dramatically steep.

The performance and reliability of a water supply for firefighting is best determined by a recent water flow test. It is generally understood that all flow tests can be regarded as being reasonably accurate, at least within 5%. The hydraulic calculation then determines if the supply is adequate for the building and its inhabitants. It is the responsibility of both the designer and the local fire service to ensure that water supply duration is also sufficient. All this information needs to be validated and is of utmost concern to property owners, insurance authorities, and primarily the fire department. It has a significant impact on safe and effective fireground operations.

Underground Piping

It would be a fabulous understatement to say to someone that the private water main supplying a fire sprinkler system is a vital link to fire suppression safety. It is so vital, in fact, that insurance carriers have for many years deducted a point charge when rating structures for fire protection quality if they did not have two separate water sources in the building. Only a very few buildings, obviously, are able to avoid this (miniscule) rate charge. But it bears mention that many of these same insurance carriers often mandate a maximum number of systems to be supplied by one underground water main. Whether that number is four, five, or six, the recommended practice is to feed the building with as many underground supplies as possible to avoid the inclusion of a header manifold with a plethora of risers, the reasoning being so that a single underground break will not sabotage the integrity of all systems protecting a very large facility.

You'll find much of what you need to know regarding NFPA requirements with regard to underground piping in Chapter 10 of NFPA #13. This chapter, which also appears in NFPA #24 (*Installation of Private Fire Service Mains and Their Appurtenances*), covers regulations for materials, pipe lining, installation, and testing. One particular section (10.6.1) often ignored or overlooked by engineers simply states that "Pipe shall not be run under buildings." If the piping absolutely must be routed under a building, the section goes on to require specific precautions, such as arching the foundation wall above the pipe, and including (indicating) isolation valves on either end of the building-buried pipe. See Figure 7.2.

The rest of the story is contained in Chapter 15 of NFPA #13. This short chapter provides the framework for sizing the water supply, and opens with "No pipe smaller than 6" shall be installed as a private service main." As usual there are caveats, which all must be met if a water supply smaller than 6 in. is desired to serve a commercial building. For example, if the building's sprinkler system is to be serviced by a 4-in. line, it cannot also service a (Class I or III) standpipe system, the hydraulic calculations must bear out

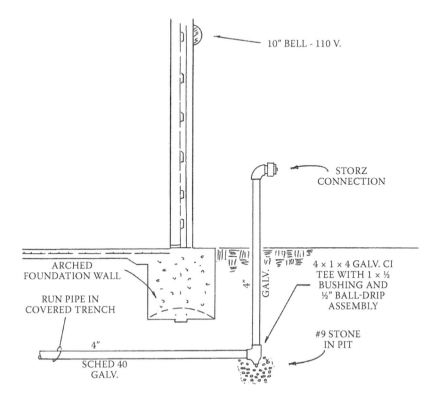

FIGURE 7.2
This underground piping originates at the fire department connection and ties into the fire sprinkler system riser. It is designed to receive water from a fire hydrant or fire engine pumper.

the adequacy of the underground pipe size(s), the underground pipe must be sized at least as large as the system riser, and the private main must be devoid of any fire hydrants.

There are numerous material options for underground pipe, however, ductile iron piping is the norm for the larger sizes used to supply automatic fire sprinkler systems. Really odd sizes are 2-1/2 in. and 3 in., but if they are called for then your best bet is to go with ductile iron if it is 3 in., and copper (Type K) for 2-1/2 in. and smaller. If permitted by local authority, underground plastic (C900) pipe can be used which, like copper, carries a C-factor of 150.

How deep you bury the pipe depends on geographic location. For the United States and Canada, you will need to reference the map charted in the appendix of NFPA #24 to determine minimum depth of cover. For example, this minimum depth would be 4-1/2 ft for Indianapolis, 5-1/2 ft for Chicago, 7-1/2 ft for Minneapolis, 8 ft for Saskatoon, and so forth. In situations where frost penetration and surface loads are not a factor, the trench excavation should be deep enough to provide at least 30 in. of cover at all locations above the pipe. The trench width should be at least a foot greater than the O.D. of the piping.

The use of underground PVC pipe and fittings has become more prevalent in recent years, partly due to its fast-track nature and its versatility. When curving the line to avoid obstructions or to make slight turns, the curvature chart In Table 7.1 outlines radius limitations. There is virtually no deflection at the joint; these curvatures only reflect the maximum bending of the pipe lengths.

Versatility and ease of fabrication are qualities also possessed by underground grooved piping that cannot be matched by flanged or tie-rod restrained systems. The use of Victaulic Style 31 ductile iron couplings allows the self-restrained joints to expand, contract, and deflect, which can counteract the stresses caused by thermal changes and settlements that can bring on-ground shifting. Style 307 AWWA-to-steel transition couplings are used for connection to the (inside) steel piping.

Thrust blocking may not be necessary when using underground grooved piping, but it is an NFPA #24 requirement for effectively anchoring fire mains.

TABLE 7.1

Pipe Curvature Chart

Pipe Diameter (in.)	Offset Length (Ft per 20 ft)
3	2.27
4	1.77
6	1.21
8	0.93

Thrust blocks serve as moorings between the fittings and the solid trench wall, and consist of a concrete mixture made with cement, sand, and gravel. Their purpose is to prevent pipe movement wherever there is a diameter or directional change in the pipe, or at dead ends. The bearing surface must be in a direct line with the major force created by the pipe or fitting.

The flushing of underground piping is a critical ingredient of any fire sprinkler system installation process. As it is the nature of sprinkler system piping to progressively diminish in size, an object (gravel, tools, stones, chunks of wood, you name it) entering the sprinkler system from underground piping will eventually lodge itself at a point where it will quickly become a major impediment to the flow of water. The discovery of foreign materials in sprinkler systems that originated from the underground water supply is not at all uncommon. This underlies the reasoning behind the code requirement that all underground flushing must be continued until the water is clear.[1]

Tanks

In rural communities and outlying suburban areas that are not annexed to any particular town, stand-alone water supplies for fire protection purposes are quite common. In many locations, a private fire protection water supply must take into account onsite water flows as well as fire sprinkler water flow. There are many methods to accomplish this task, and they take the form of enclosed tanks, elevated tanks, buried tanks, and open reservoirs. The latter would require a large chunk of land to be used for a reservoir or embankment-supported coated-fabric suction tank. If not fed by gravity, the tank will be accompanied by a fire pump situated beneath the bottom of the tank or reservoir. In many cases, the tank may be buried beneath the building. Or, a tank may simply be constructed, in any of a variety of shapes and depths, out of concrete walls above which (on the floor of the lowest level) will be a vertical turbine pump complete with a stainless steel shaft that extends down into the tank water, well, or wet pit. A basket suction strainer is affixed to the bottom of the shaft or column pipe.

The water storage tank is the most important component requiring inspection and maintenance in any facility. The NFPA #25 standard outlines requirements regarding frequency of tank testing, inspection, and cleaning. It mandates that the interior of steel tanks without corrosion protection be inspected every three years, and the interior of other tanks be inspected once every five years. The reference there is primarily to fiberglass tanks, which are lighter and rustproof, yet very durable. Routine inspections are very important issues, as are tank heating. Section 2.3.16 of the *FM Global Property*

Loss Prevention Data Sheets includes (in part) this weekly advisory: "Report if any of the following conditions are not being met: (a) full; (b) heating system in use; and (c) adequacy of temperature at cold water return. Ensure that the water temperature is maintained at 42 degrees or higher." Although there are various heat system options available, insulating the tank exterior is the most economically viable.

All water is drained from the tank for a tank inspection so that inspection personnel may enter the tank and conduct a thorough internal examination. As an alternative to this "dry" inspection, a remote operated vehicle (ROV) may be used, which does not require any draining and fully meets NFPA inspection requirements. At least monthly, tanks must be inspected to ensure that the water level and the condition of the water supply are suitable. Also monthly, the following shall be inspected:

- The tank exterior, supporting structure, ladders, or catwalks.
- The tank and support are free from a build-up of ice.
- The tank exterior area is free of combustible storage or trash.

At least twice per year, sediment shall be drained or flushed from the tank. Tank vents shall be cleaned annually, and all tank drain valves must be fully opened (and closed) at least once per year. When low water temperature alarms are provided, those are to be tested at least once per month.

Gravity Tanks

A longstanding safety feature that is much revered is the great American gravity tank. For years it has graced urban skylines, housing water and harnessing an energy known as simple gravity. Nowhere else are these tanks so prominent or seen in such high concentration as they are in New York City, where they have fed standpipes and fire sprinklers since the 1920s. Thousands remain today in the Big Apple. Prior to the proliferation of fire pumps, the wooden rooftop water tanks were long considered the first line of defense against high-rise building fires. They ensured both an adequate quantity of water and adequate pressure. Many tanks go unseen by passersby; they are housed inside the building on its uppermost floor or penthouse. Water pumped up to fill the tanks is gravity fed down through 6-in. or 8-in. pipe, normally to a shutoff valve in the basement and then upwards to the standpipe that feeds fire hose valves and sprinkler systems. A pressure tank in a heated area is used as a backup. The first fire sprinkler that flows is always pressurized. This simple methodology remains the most efficient means to supply consistent water pressure. Once the tank is filled it offers fire protection without interruption due to power failure or mechanical breakdown.

In remote areas where water is not accessible, tanks fed from wells stand as a dependable water source.

In colder climates, what keeps these enclosed tanks from freezing is its large mass of water and any small flows that occur. Exposed connecting pipe that is vulnerable to the atmosphere may require heat tracing. An annual inspection during the autumn months should include a tank examination to determine if adequate protection against freezing has been maintained and that any heating system employed is still in peak operating condition. Although wood doesn't crack in the event of a freeze-up, pipe and fittings will. It is imperative that heating equipment allow reliable operation to 50°F. Any overheating can damage a wooden tank. The exterior access ladder should be periodically checked for stability, and the NFPA requires a five-year check of the tank interior for any possible collection of debris. In areas subject to seismic activity, the tanks are to be designed for earthquake protection.

A unique quality of wooden gravity tanks is their inherent thermal ability to protect water from freezing. Natural and noncorrosive, the cellular makeup of wood serves as an efficient insulator. Cedar planks are often used for construction of the tank walls and floors because of their resistance to mold and rot. The planks, normally 2-5/8-in. thick, have the insulation equivalent of a 28-in. concrete wall. Some of the older tanks were constructed with 3-1/2-in. thick walls. These tanks, taken down after 100 years of service, have shown little (1/16 in.) or no evidence of rotting. See Figure 7.3.

Gravity tanks are extremely heavy, supported by steel towers above strong structural columns built around the building elevators. The wall construction of the tanks consists of wooden "staves" that are between 15 and 20 feet long. They are banded together with tightened 1/2-in. steel rods or "hoops," a process similar to wooden barrel making. There is no glue; leakage is prevented by the pressure exerted by the galvanized hoops combined with the swelling properties of the wood when wet (as in an old boat). A double-sided metal sign adorns each tank which provides information such as the manufacturer, date constructed, tower height in feet, tank diameter, stave length, and gallon capacity. Gravity tanks range in capacity from 5,000 to 75,000 gallons. If built today, a 10,000-gallon wooden tank will cost in the vicinity of $30,000.

Gravity tanks dot rooftops in virtually any large industrial city (and some rural towns) east of the Mississippi. Some that can be seen today are actually "dead soldiers": just a steel stand with an empty tank. Many are used as billboards, but they do present a safety hazard if not kept full of water. Rotting becomes a problem if they are dry. If they remain in place, varnish and wood preservatives are used and towers must be beefed up structurally. The steel hoops must be nailed in place to prevent them from falling as the wooden staves begin to shrink.

Cedar is a readily available wood. Almost any pencil you pick up is made of cedar. But most gravity tanks were constructed of redwood (from coastal California or China) or cypress (from U.S. southern swamps). The taller redwoods have a longer cell structure than the cypress, which has harder

FIGURE 7.3
A gravity tank in an urban landscape still in operation, serving fire sprinkler and standpipe system needs.

fiber and a longer life as a tank, easily lasting 100 years. Wood from these trees represents the highest grade of wood, which today is known as "tank grade." This wood has no knots, and comes from old, slow-growth trees with long trunks, some of which spanned 10 feet in diameter. You cannot count the trunk rings because they are so close together. When gravity tanks are removed (roughly a $5,000 endeavor), the remaining wood won't go to a landfill because it is in hot demand from woodcarvers and sign makers. They will pay up to $3,000 for the high-grade wood from a single tank.

There are about 200 gravity tanks still servicing Chicago buildings, and fire sprinkler contractors can expect a call when there's a problem. These are typically freeze-ups in the 2-in. filler line, or a circulating hot water line that didn't get drained. Or an ice plug may form in the service riser. Or a rupture will occur between the roof and the underside of the tank if a heater failed to run, and the result is a pretty big mess. But all is repairable, and a simple service call is all it takes for the entire system to be back up and running, and that is big news in Chicago. In 2006, their city council unanimously passed an ordinance to keep gravity-fed wooden water tanks from being haphazardly torn down. The ordinance imposes a 90-day demolition delay on all gravity tanks, a timeframe which the city can extend at its discretion. According to John Russick, curator of architecture at the Chicago History Museum, "So much of our expanded view of history and what's important about the way people lived is tied to ordinary objects like these tanks. That the tanks are part of the landscape makes them worth saving. Most people probably think they're still a dime a dozen, and they're not." New York has made no such rescue efforts. The spokeswoman for the NYC Landmarks Preservation Commission, Elisabeth DeBourbon, has stated, "We only designate entire buildings as landmarks. We don't designate features of buildings."

For combination sprinkler/standpipe systems, water pressure (provided entirely by gravity) is determined by the height of the tank above the particular floor of sprinklers. If you need 15 psi at the highest fire sprinkler, then the tank's water level must be 35 feet above that sprinkler. The gravity tank is vented and requires no amenities other than a level switch or float valve inside the tank which sends a signal to a domestic pump in the basement to lift water into the tank until it is filled. The greater the pump capacity is, the shorter the time needed to refill the pump. A second float valve near the top of the tank serves as the domestic pump shutoff. The tank also contains a roof hatch with access to an inside maintenance ladder. Installation of new tanks must conform to requirements outlined in NFPA #22, *Standard for Water Tanks for Private Protection*. This pamphlet includes specifications and unit stresses for the steel supporting towers as well.

In addition to fire pumps, what has put an end to new installations of gravity tanks are concerns regarding maintenance, structural integrity, and the reluctance of employers to have anyone working at high elevations. No one, it seems, wants a new one. But although they appear to be artifacts of a bygone era, those still in service provide an essential fire protection and life safety service. In New York City, assuredly the last vestige for gravity tanks, it is estimated that wooden tanks are present in 90% of structures over six stories high. But they won't last forever. The redwood tanks, particularly those harvested of wood from China's east coast, have a projected 50-year life expectancy. Many of those were erected in the 1950s and will be the next to go. And they will be missed. They've had a nice long run.

Endnote

1. Mark Bromann, "What Lies Beneath," *PM Engineer,* May 2005, pp. 36, 38, 40.

8

Commercial Fire Inspections

The goal of any fire inspection is to ensure that all building occupants, and the contents of the structure, are protected from preventable fire hazards. The inspector, to the best of his or her ability, should be able to provide a report that is proof positive that all fire protection equipment is in place and in proper condition to minimize all risk to property and human life should a fire occur. Through the exercise of proper precautionary measures, the outbreak of fire itself is often preventable. For certain, the fire safety inspector is concerned with validating that all conditions within a structure meet with the requirements of that state's fire regulations codes, local (village, town, or city) codes, and the applicable national standard fire codes.

The codes published and regularly updated by the National Fire Protection Association (NFPA) that the inspector will most often reference include the following:

NFPA #10, *Standard for Portable Fire Extinguishers*

NFPA #13, *Standard for the Installation of Sprinkler Systems*

NFPA #14, *Standard for the Installation of Standpipes and Hose Systems*

NFPA #20, *Standard for the Installation of Stationary Pumps for Fire Protection*

NFPA #22, *Standard for Water Tanks for Private Fire Protection*

NFPA #25, *Standard for the Inspection, Testing, and Maintenance of Water-Based Fire Protection Systems*

NFPA #72, *National Fire Alarm and Signaling Code*

NFPA #101, *Life Safety Code*

NFPA #25 in particular is a handy reference for all fire inspectors and should constitute mandatory reading material. The current (2008) edition covers just 98 pages, and is about as complete a document as a building engineer could reasonably expect. Requirements, recommendations, and procedures are outlined in cogent fashion with regard to fire sprinkler systems and related fire suppression equipment and systems including fire pumps, underground piping, standpipes, water mist systems, and water storage tanks. Recommended intervals for various maintenance and testing procedures of all fire protection equipment are organized in easy-to-read tables. Of note, Section 3.3.18 defines an inspection as one that consists of "A visual examination of a system or portion thereof to verify that it appears to be in operating condition and is free of physical damage." With that being said, the inspector's responsibility is limited to a walk-through visual inspection from the floor. That is not to say that the inspector cannot climb a ladder to examine a space above a ceiling with a flashlight. Many will do just that, but this type of activity is not necessarily a required duty of the inspector.

At a seminar in Middlesex County, New Jersey, I asked a roomful of code officials what the most common sprinkler system deficiency was that they noted during a typical fire inspection. The number one response was painted sprinkler heads. Painting is the primary, but not the only cause of a problem known as loading, which occurs when a buildup of foreign material on the sprinkler delays or prevents proper response of the sprinkler in a fire condition. Sprinklers must be replaced if the loading cannot be easily removed from the sprinkler and its fusible elements.

Another common problem that an inspector runs across is storage items piled up around sprinkler heads. Spray obstructions can prevent water discharge from reaching the actual seat of the fire. For that reason, the NFPA #13 standard requires a minimum of 18 in. between the sprinkler deflector and the top of storage. Concerning performance objectives for sprinklers, the sprinkler pattern will equal an 8-ft diameter circle at a point of elevation 18 in. beneath the sprinkler head,[1] given a standard half-inch sprinkler with a 7 psi residual water flow.

The code officials pointed out to me that they will routinely include the existing fire department connection in their inspection. Specifically, the connection must be visible, situated in a spot that is reasonably accessible to a nearby live city hydrant, and free from interferences by fences or landscaping. NFPA codes recommend that the height of the connection be between 18 and 48 in. above grade. Of course, it is important that the connection's threads match those of the local fire department. All of this is crucial mainly because the fire department connection quickly becomes the sprinkler system's sole water supply in the event of an accidental valve closure.

A commercial fire inspector should have a handle on spacing and location requirements for fire sprinklers, which are discussed in great detail in Chapter 8 of NFPA #13. He must look for unprotected areas or rooms, and other sprinkler protection problems that have arisen due to the relocation or addition of walls during building remodeling. Fire sprinklers must be located a minimum of 4 in. from any wall and positioned at least 6 ft apart from each other in a single room. Sprinklers must be installed beneath ductwork (and similar horizontal obstructions) greater than 4 ft in width. Sprinklers must be installed at a location where heat is quick to collect, so that the sprinkler can fuse. In general, for flat ceilings, this means that the sprinkler deflector should be within 12 in. of the ceiling. Sprinklers installed at lower elevations have poor response times which may result in critical delays of water application.

Sometimes an upright sprinkler will be spotted that has been installed in a (downward) pendent position. The resultant spray pattern of this configuration is inefficient, and the sprinkler will need to be replaced with a true pendent sprinkler possessing a notched deflector. The same problem has been noted in cases where an "upright sidewall" needs to be replaced with a "pendent sidewall" sprinkler. In very old systems, inspectors still see 3/4-in. steel piping hung in place at the end of branchlines. Because the use of 3/4-in. steel pipe was phased out in the 1940s, it is imperative that the building owner replace this old piping (which is by now likely clogged with buildup and corrosion) with new 1-in. pipe and fittings.

A fire inspector or building owner can best be educated and prepared by studying reports of past fires in sprinklered buildings. The leading cause of sprinkler system malfunction has taken the form of partially or wholly closed system control valves. An alert inspector will look not only for closed valves, but also existing unsupervised valves. It is an important safety measure to either have a tamper switch installed on a control valve, or to have the valve securely chained in the open position, particularly when the valve is concealed above a ceiling or is otherwise unseen by the building engineering staff. To answer the question, "Is this building completely sprinklered?" an inspector should be on the watch for unsprinklered, combustible, concealed spaces. Fire can linger, smolder, and spread behind walls and above ceilings or anywhere that combustible construction exists without sprinkler protection. History can confirm that inspectors have often missed this lethal "time-bomb" on routine inspections. Every seasoned member of the fire protection community is familiar with cases of large-scale fire losses that were attributed to the spread of fire in nonsprinklered areas.

The first question that the inspector asks himself before an inspection is, "What is particular to this occupancy?" From business to business, the type and level of fire protection provided varies from safety need to safety need. Fire safety needs are relative and will be vastly different if the occupancy is a restaurant as compared to a bank, printing plant, or a high-piled storage warehouse. Code requirements differ depending on occupancy, on building size, and on building elevation. This is essential for fire risk assessment and overall fire prevention. The inspector must be able to classify occupancies in order to recognize whether the existing suppression systems were designed in a fashion to protect the in-place tenant activities. The aforementioned code officials correctly reported that the number one fire protection enemy they fear is a change in occupancy. An old pipe-scheduled system, for example, is without question deficient and lacking in capacity to control a fire that could occur in an area containing storage of commodities piled to a height exceeding 12 ft.[2]

One reason that in-house inspections are necessary and integral to business as a whole, is that they make the plant safety manager a visible figure in all facility operations. The places where losses can occur are where the plant inspector should be spending the lion's share of her or his time, giving a sure indication to employees that loss control is a priority.

When called on to witness tests, the inspector can do a building owner a favor by paying close attention to particulars and diverse conditions. During a fire pump test, for example, it should be verified that there is a bypass installed around an electric-drive fire pump. She may notice that the fire pump room itself is without sprinkler protection. Or it may contain a partially closed valve, or exposed electrical wiring. Or there may not be a stock of spare sprinklers available. The municipal official or commercial inspector is well within her or his scope of work to note that such conditions contribute to reduced system integrity.

Testing of the main drain on each fire sprinkler system involves opening the 2-in. angle valve on the system riser. The point of this test is to detect any possible obstructions in underground feed piping. To be on the alert for this, the inspector needs to look first for a considerable pressure drop when the drain is fully opened. A problem may also exist if it takes an inordinate amount of time for normal pressure to return once the angle valve is (slowly) closed. Either of these occurrences could be symptomatic of some sort of obstruction in the underground pipe or a valve that is closed in the street. Properly working and accurate pressure gauges are necessary for this testing. NFPA #25 recommends that gauges be inspected monthly and also that they be either tested or replaced at five-year intervals. Manufacturers predict a pressure gauge "design-life" to be roughly 12 years.

Procedures mandated for fire pump tests are outlined in Chapter 8 of NFPA #25. Because of the high water pressures involved with discharging water during this testing, safety should come first. The firm conducting the fire pump test should ideally be a company specializing in performing this type of work, with a minimum of three years documented experience and quality approved by the fire pump manufacturer. The fire pump must be tested on an annual basis. In addition, NFPA #25 requires that the pump be tested weekly, without flowing water, by starting the pump automatically. If the pump is driven by an electric motor, it is to be run for a minimum of 10 minutes. Any fire pump that has been submerged to any extent in floodwaters must be immediately inspected and tested.

A common question from inspectors is whether the fire pump should be shut down when conducting a 2-in. main drain test. There are two schools of thought on this. If the inspector desires completely normal conditions, then the drain test should be performed with the pump on. However, in the case of a partially closed valve in the municipal underground piping network, the subsequent running of the fire pump can mask a problem. The greatest likelihood of detecting the condition of a partial valve closure is with the pump off.

System test requirements are also outlined in the BOCA (Building Officials and Code Administrators) National Fire Prevention Code. All fire protection systems (with the exception of smoke control systems) must be tested for performance every 12 months. The tests must be conducted by a representative of the owner in the presence of the code official or authority having jurisdiction. A complete written record of all required tests must be maintained on the premises.

Testing of fire detection and alarm systems must also follow requirements dictated by the recognized fire codes. During the inspection and testing of fire alarm systems, the inspector will be mindful of differing safety mechanisms within the building. As examples, he will check to see that fire extinguishers are situated in locations that are accessible and reasonably close to hazardous areas. He will check building exits to verify that nothing is blocking the exits, that doors can be easily opened, and that there are enough exits in the building in relation to the occupant load. He will also check for any highly flammable substances within the building and may recommend an alternate location for their storage, safely away from plant areas of high risk. Naturally, all building occupants must be put on alert that a fire alarm test is forthcoming, so that chaos will not ensue. It may prove optimal to schedule a fire drill coinciding with the fire alarm test and inspection. All personnel inside a building in a working environment need to know exactly where to go in the event of a fire. For reasons besides just the avoidance of false alarms, the systems will be frequently inspected. Even the most technologically advanced fire alarm systems will not work for a prolonged period of time if they have been installed incorrectly. Section 5.2.6 of NFPA #25 reads that "Alarm devices shall be inspected quarterly to verify that they are free of physical damage."

Although certainly not an etched-in-stone requirement, conversations with the building engineer or her staff are paramount in turning up "red-flag" maintenance problems. Being informed of frequent dry-valve tripping or the constant starting and stopping of a dry-pipe system air compressor may lead the inspector to discover the existence of leaks in the sprinkler system. Or it may be that dry-valve pressure settings are simply improper for real conditions, as would be the case when a dry valve accidentally trips during a fire pump test.

Inspectors have the opportunity to be certified by NICET (National Institute for Certification in Engineering) through written testing and by the validation of actual work experience. The NICET subfield "Inspection and Testing of Water-Based Fire Protection Systems" includes an examination covering over 70 separate work elements. For this, candidates will need to familiarize themselves with requirements of NFPA #25 that may be considered less than casual knowledge: for example (Section A.13.3.3.2), to prevent jamming of a wall-post indicator valve, the valve wheel in the open position should be backed one-quarter turn at regular inspection intervals. And (Section 13.6.1.2), the weekly inspection of a reduced-pressure principle backflow preventer shall include an acknowledgment that the relief port is not discharging at a high frequency. Also (Section A.5.3.4), within an antifreeze loop, listed CPVC sprinkler pipe and fittings should be protected from freezing with glycerin only. The use of diethylene, ethylene, or propylene glycol is specifically prohibited. When inspecting antifreeze systems employing listed CPVC piping, the solution should be verified to be glycerin based. Many refractometers are calibrated for a single type of antifreeze solution and will not provide accurate readings for the other types of solutions.

As mentioned earlier in this chapter, NFPA #25 is considered required reading by any code official involved with both system inspections and the witnessing of system tests. By intermittently leafing through that standard, a fire inspector will invariably come across numerous code requirements of which most plant engineers are completely unaware. Inspectors may take the form of safety managers, fire prevention bureau officials, fire inspection firms, or fire sprinkler contractors. Regardless of who inspects existing commercial facilities, the work should not be regarded as something representative of more restrictive fire protection requirements, but rather as common-sense fire safety.[3] Each involves many facets common to life safety practices. Failure to follow policy or procedure is a recipe for real trouble, and this is all the motivation an inspector needs to carry out the job.

Endnotes

1. The NFPA 13 code section to reference for this mandate is 8.5.6.1, however, it should be duly noted that *FM Global Property Loss Prevention Date Sheet* 2-0, in Section 2.2.2.1 and under Section 2.1.2.1 regarding clearance below sprinklers, advises to "maintain a minimum 3 ft clearance between the deflector of a sprinkler and any combustibles located below it."
2. Mark Bromann, "Sprinkler Inspections," *PM Engineer,* April 1998, pp. 30, 32, 33.
3. Mark Bromann, "Sprinkler Inspections," *PM Engineer,* May 1998, pp. 24, 27.

9

Fire Alarms

Most facilities today are equipped with fire alarm systems, however, their components and conceptual integration have been created by different manufacturers and as such, may employ differing protection mechanisms. Because of this, the building owner must retain a set of installation drawings, operation manuals, and a detailed sequence of operation for the particular fire alarm system that he owns. We rely daily on well-engineered alarm monitoring systems that will notify fire departments of a fire condition within seconds of the operation of an alarm-notifying device. Fire alarm systems must be installed in accordance with the requirements contained in (National Fire Protection Association) NFPA #72, the *National Fire Alarm and Signaling Code*, and in NFPA #70, the *National Electrical Code*. All fire alarm systems and equipment must be well-maintained, inspected regularly, and returned to normal after each test and alarm.

Depending on the hazard being protected, a fire alarm system includes one or more of several types of alarm-initiating devices. One of these types that everyone is familiar with is the manual pull-station. These are basically electrical switches, and most can be reset after use. NFPA 72 mandates that manual fire alarm boxes be conspicuous, unobstructed, and accessible. Designers typically locate a manual pull-station, with the lever at a height of 50 inches above the floor, at each building exit. Additional pull-stations must be provided so that there is a maximum travel distance of 200 feet (61 meters) to the nearest manual fire alarm box measured horizontally on the same floor.

Heat Detectors

The type of detection device to be implemented normally depends on design decisions made by the consulting engineer prior to construction of a new building. For example, a fire alarm designer will typically call for automatic heat detectors to work in supplemental coordination with a preaction or deluge system. For those system types, the valve controlling water discharge to sprinkler nozzles will not open until a signal has been received from the heat detection device.

The simplest type of heat detector is the fixed-temperature unit. Heat from a fire activates this detector, which contains a fusible alloy that melts rapidly at a specific ceiling temperature. This action causes the internal electrical contacts to operate, which in turn initiates the alarm signal. Temperature settings for these devices will vary (just as with sprinkler heads), with 140°F being a good call for normal conditions.

NFPA #72 specifies coverages and areas of application for the installation of heat detectors that depend on a variety of factors including ceiling height. The coverage area per detector is maximized when there is a smooth ceiling with no airflow obstruction. Very high ceiling heights and ceiling obstructions are two common challenges posed to the designer of a fire alarm system, and she or he must be familiar with the parameters called out in Section 5.6.5.5 of NFPA #72, which apply to required heat detector linear spacing when high ceilings are encountered. It's also important to bear in mind that the inevitably needed maintenance of detection systems is sometimes overlooked when the devices have been installed at high elevations.

A rate-of-rise heat detector includes a small air vent that allows for expansion and contraction of air inside the unit's housing. When temperature changes in the room cause the air inside the device to expand more quickly than it can escape from the air hole, pressure is exerted on the diaphragm, which makes the electrical contact. As is the fixed-temperature heat detector, this device is reliable and has a low false-alarm incident rate.

The advantage of the rate-of-rise heat detector is that, even if the space above the point of fire origin has not yet reached 140°F at the ceiling, it will initiate an alarm as soon as the rate of temperature rise exceeds a predetermined value, such as 12°F per minute. There is a special label on all detectors classified for hazardous areas. Explosion-proof wiring is required for any detector located in an electrical equipment room.

Smoke Detectors

Smoke detectors are the less expensive option, often featured as the "blue light special" at K-Mart. Available since 1960, these devices detect fire by sensing "particles of combustion" and certain gases in smoke. With an ionization smoke detector, the design intent is that they are "on" at all times. Ionized air in the sensing chamber serves as a conductor between two charged electrodes. When smoke particles present themselves, the air conductance is decreased. When decreased sufficiently, the electrical charge is "off," and that quickly signals an alarm.

The absolute maximum spacing recommendations for smoke detectors is 900 sq ft (and not exceeding 41 ft apart) and less than this when an increased response time is desired. Ionization detectors will give off a false alarm when humidity exceeds a high level, such as 90% or more. Don't have these installed next to a shower door.

A photoelectric smoke detector, considerably more expensive, contains a photosensitive device. It doesn't really "see" the smoke, but operates when there is a change in the intensity of light caused by the presence of smoke. Another type of photoelectric smoke detector contains a displaced photosensitive receiver, which only initiates an alarm when the smoke particles collecting in the chamber scatter the incoming light, causing light from the source to then strike the receiver. Both unit models have been proven to be very effective in spaces with a potential for smoldering fires.

Inasmuch as smoke dissipates less rapidly than heat, smoke detectors are generally a better design option for larger open rooms. And measurable amounts of smoke almost always precede measurable amounts of heat, which is one reason why engineers are compelled to call for both sprinklers and smoke detectors inside a building. Computer rooms may be protected with a combination of photoelectric and ionization smoke detectors, the principal concept being that both have to operate before the alarm system is activated. That way, alarms will not sound until there is surely a fire condition.

Smoke detectors have been a source of annoying false alarm problems for decades, although nuisance alarms received from newer commercial alarm systems installed with improved technology are becoming increasingly uncommon. The NFPA reports that only 8% of smoke alarm failures in the last several years have been due to hardwired power source problems, including disconnected smoke alarms, power outages, and power shut-offs. Hardwired smoke alarms are typically interconnected, so that if one sounds, they will all sound. Although they have an excellent performance record, NFPA statistics show that battery-powered smoke alarms have operated successfully only 75% of the time in actual fires considered to be large enough for detector activation.

Flame Detectors

Another detection component famous for false alarms is the very sensitive, but very quick-responding, flame detector. Usually you will find these detectors in occupancies with aerosol-filling or paint-mixing operations, or other areas where flammable vapors or combustible dusts are present. "Optical"-type flame detectors are expensive but, due to their superior speed of detection, they are essential for rooms with explosion potential. They are able to sense sparks, flames, glowing embers, or other sources of radiation. Of the two types, ultraviolet and infrared, the ultraviolet detector is the fastest to respond, making it an ideal choice in a hazardous location where very rapid fires may occur. The field placement of these devices is critical for success, and care must be exercised so that large equipment or piled storage does not interfere with their "line of sight." Because of the potential of flammable vapors igniting, automotive plants will use high-speed optical detectors in areas where they apply paint.

Other Detector Types

The detectors described above are the most commonly purchased, however, there are numerous other types reserved for use in special occupancies requiring differing levels of detection or maintenance. Laser detectors are reliable devices used when a very high level of sensitivity is desired. Filter detectors can be installed in very wet or very dusty environments. Line-type detectors are used most frequently in freezer warehouses. The many different types of gas detectors that are manufactured are designed to detect leaks of specific types of gases. For example, CO_2 detectors warn of excessive levels of carbon monoxide by measuring the opacity of the light in a specific area to determine exactly how much toxic exhaust is present. Power plants use combustible gas detectors quite extensively. The operation of any of these detectors can be made to activate an accompanying alarm or extinguishing system.

Alarm Systems

The most essential ingredient of any fire alarm system is continuous monitoring. Integrated with the alarm-initiating devices, the alarm system is a life safety building feature that simultaneously alerts the fire department while loudly notifying building occupants of the immediate need to evacuate the structure.

Fire alarm systems come in many varieties: six to be exact. Two of these, local fire alarm systems and emergency voice/alarm communication systems, will at minimum sound a local alarm to notify building occupants of a fire. With emergency voice/alarm communication, the occupants also receive information along with instructions pertaining to the emergency at hand. Both of these system types may also provide prompt notification to the local fire service.

In a fire emergency, elevators are simply unsafe. Building occupants may hit a button for an elevator that never arrives, causing them to wait instead of rushing to safety. If an elevator did arrive, it may become overcrowded to the point that the doors could not close, and an ensuing state of panic may erupt into chaos. A power failure caused by the fire could trap passengers inside an elevator. Because of these and other potential life-threatening dangers, the *Elevator Safety Code* mandates automatic elevator recall, whereby a fire signal from the control panel will simply direct all elevators to the first floor, opening the door if the elevator "thinks" the floor is safe. However, doors will not open on the first floor if detection devices are "telling" the panel that the fire exists on the first floor in the vicinity of the elevator door(s). Instead, it will open the doors on the next highest floor, allowing any passengers to exit, and then subsequently discontinue its service.

The alarms from an auxiliary fire alarm system are received by a municipal signaling system for transmission to the public fire service communication system. All protected properties within that community have a similar connection to the same alarm protective signaling system. This setup is in contrast to proprietary station systems, which are in wide use at very large industrial facilities containing various properties under one ownership. With that system, transmitted signals are received and recorded automatically at a supervisory station that is situated at either the protected premises or at another location of the property owner.

The central station fire alarm system consists of the operation of circuits and devices transmitted to a manned central station operated by an independent firm. That firm is in the business of monitoring and maintaining signaling systems and retransmitting alarm signals it receives to the appropriate public fire service. Except for the means of electrical signal transmission, this is a lot like the proprietary system. What is similar in scope is the remote station signaling system, which transmits signals from one or more premises to an off-site remote monitoring location (approved by local authorities), where the appropriate action is taken. The signal is transmitted over a leased telephone line to the police station, fire department, or telephone answering service. If the remote station is at the police station or the answering service, fire department personnel will receive instant notification.

All fire alarm systems have alarm-initiating device circuits that interconnect the fire alarm control panel to automatic detectors, pull stations, tamper switches, and fire sprinkler system water flow indicators (flow switches). Another feature that all fire alarm systems share is that they are each required by code to have a backup source of electrical supply.

Requirements for all systems and system installations are covered in meticulous fashion in NFPA #72 and in Articles 760, 770, and 800 of the *National Electrical Code* (NEC). It is a requirement that all of these systems have a backup standby source of electrical power. It is important that all equipment be listed and labeled by a testing and inspection agency laboratory and installed in accordance with the manufacturer's specifications. The NEC also contains requirements for properly wiring a fire alarm system to ensure peak performance, to prevent personal injury, and to prevent the wiring itself from starting a fire. Professionals rightfully point out that although conventional alarm systems have been installed in commercial buildings for many years, the newer systems keep getting better and better. Improved technology is bolstering the reliability of fire alarm systems today.

Certain information, such as the phone number of the agency that monitors the property, and the account number for the building, is required to be posted on the inside door of the control panel. As most panels have a "dual-connect" feature, the panel information should also indicate whether the signal goes to the police department or the fire department. NFPA #25 requires that the alarm-receiving facility be notified before and after conducting any test or procedure, to avoid the possibility of false alarms. The signal that the remote monitoring station receives may indicate either "fire" or "trouble." The "trouble" signal can mean some anomaly within the system itself, subscriber error, or telephone link problems. It may also indicate any number of nonfire conditions, such as low power (to an electric fire pump), low building temperature, valve position tampering, and so on. In some cases the "trouble" signal may even mean that the burglar alarm was tripped, or that the building experienced a temporary power outage. However the panel is programmed, all signals detecting undesirable events are received instantaneously and exact times are recorded.

What distinguishes the modern-day fire alarm systems are their interconnective capabilities, or "single-seat control." Beyond simply transmitting a fire alarm, they are able to activate elevator recall, control exit doors, and actuate preaction and deluge extinguishing systems. They are also able to give a firefighter information regarding the exact location of smoke and heat in a building. The receipt of a signal from one or more detection devices may also be utilized to operate the building's air-handling equipment for smoke control. Here, the fire alarm system can truly be viewed as a life-saving device, because smoke can be extremely lethal. Smoke control may be of two types: passive or active. With passive smoke control, all fans and smoke dampers (within the ductwork) are shut down during fire conditions. With active smoke control, the air-conditioning system is used to exhaust smoke from inside the building to the exterior.[1]

Inspections and Testing

The inspector, whose job it is to approve the original installation, conducts the first inspection of any fire alarm system. Inspectors should be provided with a set of as-built installation drawings, operation manuals, and a written sequence of operation. In general, inspectors are looking for an installation that is both code-conforming and representative of quality workmanship. Anything improperly or shoddily installed will be difficult to maintain, and probably fall into disrepair before long. The systems should test free of grounds, with the exception of those parts of equipment that are intentionally grounded to provide ground-fault detection, noise suppression, emergency ground signaling, or circuit protection grounding. The conditions in locations in which the apparatus is installed must not exceed certain voltage, temperature, and humidity limits. Optical fiber cables must be protected against mechanical injury. All sensor detectors must be located and mounted so that vibrations or accidental jarring will not cause them to operate. NFPA #72 also mandates that any initiating device subject to possible mechanical damage be protected by a mechanical guard that is specifically listed for the detector it protects.

Ultimately, the building owner is responsible for the ongoing inspection, testing, and maintenance of the entire system unless that is delegated to another party by means of a written contract. Inasmuch as there is very little standardization with regard to the way in which controls, icons, trouble signals, and lights are arranged from one fire control panel to another, it is imperative that several of the building owner's on-site personnel be familiar with the operational manuals for the particular alarm system that has been installed.

Fire alarm equipment must be maintained in normal operation and always returned to normal as quickly as possible after each test and alarm. If the system is going to be out of service for more than four hours, the authority having jurisdiction must be notified so that a fire watch may be provided. Continued operational reliability of the alarm system is achieved through maintenance of the equipment. Regular maintenance procedures must be developed, followed, and recorded, to avoid a "set it and forget it" mentality. Alarm and supervisory signals must be tested annually. Any lead-acid or dry-cell batteries are to be visually inspected and voltage tested on a monthly basis. Depending on the site conditions for each area, a cleaning schedule is also a good idea in order to remove dust, dirt, dead insects, and so forth, that can adversely affect appliance performance.

Reacceptance tests must be carried out after any modifications are made to the system, such as the addition or deletion of system components, repair or adjustment to the system hardware or wiring, or following any change in software. A permanent log should be kept for the recording of all inspections, testing, and maintenance activity.

Simply put, periodic testing is vital because the integrity of the systems must be maintained. There are a number of methods approved by manufacturers to ensure that smoke detector sensitivity is not compromised. Qualified personnel must visually examine the initiating device units semi-annually, and physically test all detectors annually. NFPA #72 requires that testing operations include verification that the releasing circuits and components energized by the fire alarm system are electrically supervised and operate as intended on alarm. The idea is to safeguard system continuity.

Fire alarm systems for any business operation provide a direct link to safety. That particular element of time hovering between ignition and detection has more to do with whether we are going to save property and lives from fire than any other single feature of fire protection. The fire alarm industry has long interfaced with those in the fire sprinkler industry and with the nation's municipal firefighters in a collective unity of purpose. These efforts, along with advanced mechanical and electrical technology, are continuing to bolster the effectiveness and reliability of fire alarm systems.

Endnote

1. Mark Bromann, "They Work in a Flash," *PM Engineer,* March 2001, pp. 26, 28, 31.

10

Fire Extinguishers

Fire extinguishers are an integral part of any fire protection program. They have been around for well over 100 years and are seen as a first line of defense against small fires, whether or not a fire sprinkler system also protects the building. They are required components because where you provide fire protection it would be an oversight to provide just one line of protection. In London last year, I noticed an onslaught of fire extinguishers placed in numerous commercial and retail establishments. From small shops to hair salons, their trademark red color jumped out at me. Sometimes these extinguishers were situated on some sort of colored mat, however, they typically sat right on the floor, surrounded by various objects or items for sale, and without any accompanying signage or fanfare. I asked some of the shopkeepers about the fire extinguisher kept on the floor in their place of business. "That's the law," was the typical response.

For the record, keeping the fire extinguisher on the floor of any establishment would represent a code violation in most American cities. National Fire Protection Association (NFPA) #10 requires that the bottom of the fire extinguisher be located at least 4 in. above the finished floor, and the top of the handle must be lower than 5 ft above the floor. Some towns require a specific height. Schaumburg, Illinois, for example, requires the top of the extinguisher's handle to be set exactly 42 in. above the floor. Most fire departments like to see the extinguisher installed about waist-high, and higher if children are likely to be around.

Few people have ever heard of George Manby, but he is credited as the inventor of the portable fire extinguisher, in Virginia in 1813. The following quote from George Manby regarding his invention also qualifies him as something of a forefather of fire protection engineering: "A small quantity of water, well directed and early applied, will accomplish what, probably, no quantity would effect at a later period." The first actual U.S. patent for a fire extinguisher was issued in 1863.

Although recognized as a first line of defense against fire, it is important to also recognize that the portable fire extinguisher is no substitute for the fire department. Roughly one-third of all people injured by fire are hurt while trying to control it. The devices are to be used only on small blazes, and occupants should always call 911 before using any fire extinguisher. That way, if the fire cannot be controlled, the fire department will already be on their way.

An extinguisher should be mounted in an easy-to-reach and very visible location, preferably near a door or entrance, so that the user will not have

TABLE 10.1

Approximate Sizes and Range of Popular Extinguishers

Capacity	Unit Height	Shell Diameter	Typical Range of Stream
ABC ammonium phosphate based Dry Chemical Fire Extinguisher			
2-1/2 lb.	15″	3″	12 feet
5 lb.	16″	4″	15 feet
10 lb.	19″	6″	20 feet
20 lb.	24″	7″	20 feet
BC Sodium bicarbonate based Dry Chemical Fire Extinguisher			
2 lb.	14″	3″	10 feet
5 lb.	15″	4″	15 feet
10 lb.	22″	6″	20 feet
20 lb.	24″	7″	20 feet
BC or ABC portable Carbon Dioxide Fire Extinguisher			
5 lb.	17″	5″	5 feet
10 lb.	19″	7″	5 feet
15 lb.	25″	7″	5 feet
20 lb.	26″	8″	5 feet
Pressurized AFFF foam (for Class A or B) Fire Extinguisher			
2.5 gallon	25″	7″	20 feet
Class A Pressurized Water (or antifreeze) Fire Extinguisher			
1.5 gallon	20″	6″	20 feet
2.5 gallon	25″	7″	35 feet

to pass flames and smoke in order to reach it. An extinguisher is not to be mounted at the very spot of kitchen equipment; you will want to be able to grab the extinguisher from a remote location in the event of an oven fire. A very smart place to keep a home fire extinguisher is somewhere in the garage.

Portable fire extinguishers are to be used according to the instructions found on the unit's label, and please note that they contain a limited quantity of extinguishing material. The typical duration of use for a 5-lb. fire extinguisher is about 20 seconds, and only about 40 seconds for a 10-lb. appliance (see Table 10.1).

Classifications

Extinguishers are labeled either A, B, or C, depending on the type of expected fire for the corresponding occupancy. Class A fires are those in which the burning items are ordinary combustibles, such as upholstery, wood, clothing, rubber, paper, and most plastics. Class B fires are those involving flammable liquids, oils, paints, propane, and grease fires in kitchens. It is dangerous and futile to use water or Class A fire extinguishers on Class B fires, or Class C fires, which involve electricity and can include burning wires, electrical equipment, or power tools that might be "live." Fire extinguishers carrying a "BC" label can be used on both Class B (flammable liquid) and Class C (electrical) fires, but are less effective on deep-seated Class A fires.

Water extinguishers are generally silver (chrome) in color. They have a flat bottom, a long narrow hose and carry about 2-1/2 gallons for Class A fires. For some of the older models, you must turn the extinguisher upside-down to use it. It will only discharge water in an inverted position. Recent changes in the code are showing aluminum fire extinguishers the way to the retirement home, as hydrostatic testing of these units is prohibited.

A very common extinguisher is the ABC-rated multipurpose dry powder extinguisher, which is red in color and equipped either with a long narrow hose or just a short nozzle. These extinguishers are very light and portable, and filled with a dry chemical powder. In commercial settings, fire extinguisher types may contain foam, a clean agent gas, or carbon dioxide. The latter is a heavier model, ranging in weight from 15 to 85 pounds. These are also red, have a large tapered nozzle and are found in special industrial areas such as aircraft hangars and high-voltage electrical rooms. These are high-pressure cylinders, definitely not to be dropped. A carbon dioxide extinguisher discharges a stream of gas as opposed to a liquid. The gases put out the fire by displacing the oxygen from the air, thus smothering the fire. Also, by absorbing the heat of a fire rapidly it is effective on electrical fires. The major precaution to observe when using carbon dioxide extinguishers is to guard against the reignition of hot combustibles. Some manufacturers do not recommend carbon dioxide cylinders at all because they are so heavy and do not have an extended range. Dry chemical extinguishers are generally a better option for many industrial occupancies.

Of all the available extinguishing agents, only foam can secure a fire, making it the advantageous choice in the event of a flammable liquid spill. The aqueous film-forming foam (AFFF) extinguishers can be used on any fire, contain 2-1/2 gallons of solution, have an operating range of 20 ft, and a typical duration of 55 to 65 seconds. This type of extinguisher, like the Class A water base extinguishers, cannot be installed in any areas subject to freezing temperatures, as it will itself freeze. It contains either a chemically generated expellent or compressed nitrogen. The amount of foam discharged is normally about eight times the contents of the extinguisher. AFFF extinguishers should be recharged regularly, at a duration recommended by the manufacturer.

According to NFPA #10, placement of the fire extinguisher depends on the class of fire expected. The requirement for walking distance to an extinguisher is 50 ft for Class B extinguishers as opposed to 75 ft for Class A extinguishers, simply because an extinguisher must be used sooner on Class B fires. Although a Class A fire may develop slowly, the maximum intensity of a flammable liquid fire is reached very quickly. Following the same path of logic, the maximum recommended travel distance for Class C fires drops to 30 ft.

To ensure that its contents can be discharged towards a fire with sufficient force, pressure gauges were added to extinguishers in the 1950s to display the internal pressure. A tamper seal is implemented to secure the pull pin or locking device. A cardboard tag attached to each extinguisher indicates the month and year that maintenance on the unit was last completed, also identifying the firm that performed that service.

Operation

There are a few simple fundamentals involving the operation of a fire extinguisher, but the first advisory is never to even pull the pin if you cannot fight the fire with your back to an escape route. And if you run out of extinguishing agent, if the extinguisher proves to be ineffective in any fashion, or if the fire spreads beyond the spot of origin, then it's time for you to run. There is no time to gather any possessions or pets. Close the door to the room and exit the premises.

When using an extinguisher, the first thing to do (after calling 911) is to remove the plastic tie or tamper seal from the pull pin. Keep the extinguisher upright. Then use the "PASS" system, which goes as follows: *PULL* the pin, *AIM* at the base of the flames, *SQUEEZE* the handle, and *SWEEP* from side to side.

The part of the PASS system that most often goes awry is the second part of the equation. It's crucial to fight fires by aiming the nozzle low, focusing on the base of the fire. If piled items are burning, start low and then move upwards. Try not to use up all of the extinguishing agent in one single blow. It's best to start from a distance of 6 to 10 ft from the fire and then move in slowly. Stay calm and don't charge the fire. If outside, approach the fire from upwind. For a spill fire, you may have to circle around the spill. If you are encountering a flammable liquid or grease fire, you must be very careful not to splash fuel, as this will only compromise the intent of controlling the blaze and will increase the possibility of injury. In any case, sweep the extinguisher from side to side until the fire is out, and then continue to wait, as sometimes a reflash can occur.

Recharge

All rechargeable fire extinguishers are to be recharged after they have been used, even if only a small amount of agent reserve has been released. The quantity of agent to be recharged will be confirmed by weighing the unit.

The total weight of the cylinder and agent is always marked on the extinguisher label. It is a requirement that all nonrechargeable fire extinguishers be removed from service after 12 years from date of manufacture. A fire extinguisher of similar size and rating must be left in its place.

As with any mechanical device, fire extinguishers require periodic checking and maintenance. Pressurized fire extinguishers may lose their pressure over time and will not be safely operable unless they are recharged. If you are new to an existing work environment, there are five things to check in relation to the fire extinguishers:

1. Familiarize yourself with the extinguisher locations.
2. Make certain the class of extinguisher is compatible with the occupancy.
3. Ensure that the nozzle and the plastic seal on the pin are both intact.
4. Feel the weight of the unit to ascertain if it is sufficiently full.
5. Check the gauge to see that the pressure (100–175 psi) is in the "green" area.

The placement of an extinguisher should be realistically free from any possibility of mechanical damage. The most important aspect of a monthly inspection is to validate the fact that the extinguisher is actually still there. Random theft of these units, especially in large facilities, is not an uncommon occurrence. NFPA #10, *Standard for Portable Fire Extinguishers,* outlines regulations governing fire extinguishers such as distribution and spacing, classifications, selection, testing procedures, and matters of maintenance. In England, the Loss Prevention Council (LPC) offers similar publications for these purposes.

Preventing and surviving fires is a global concern. To realize this critical goal requires maintenance, inspection, and preparation. Endeavors that may seem to be redundant are necessary for a complete fire prevention effort and often, this redundancy will help to avoid the spread of fire from something small to something of tragic proportions. There is no time left to figure these things out when there is an outbreak of fire; we have to be prepared. The time to ask yourself, "Where can I find a usable fire extinguisher in here?" is before the fire occurs.[1]

Endnote

1. Mark Bromann, "Portable, Personal Protection: Just Add Water," *PM Engineer,* January 2001, pp. 29–30.

COLOR FIGURE 1.1
A firefighter checks for lingering smoke and flames following the extinguishment of an overnight fire.

COLOR FIGURE 6.4
This wet-pipe system riser includes two O, S, and Y valves with tamper switches, a double-check detector assembly, pressure gauge, flow switch, 2-in. main drain, and a spare-head cabinet mounted nearby.

COLOR FIGURE 6.6
A close-up of a dry system riser showing the dry pipe valve and trim.

COLOR FIGURE 11.2
A horizontal split-case electric fire pump.

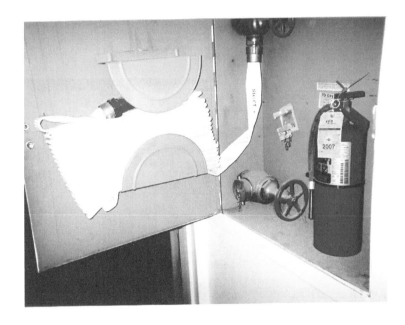

COLOR FIGURE 12.1
A Class III standpipe supplies this fire hose cabinet which houses both a 1-1/2-in. valve with lined fire hose and a 2-1/2-in. fire hose valve for use by the fire department, with room left over for a fire extinguisher.

COLOR FIGURE 19.2
A garage fire may spread to other areas of a multiresidence if not extinguished in timely fashion by firefighting personnel (photo by Steve Bittinger).

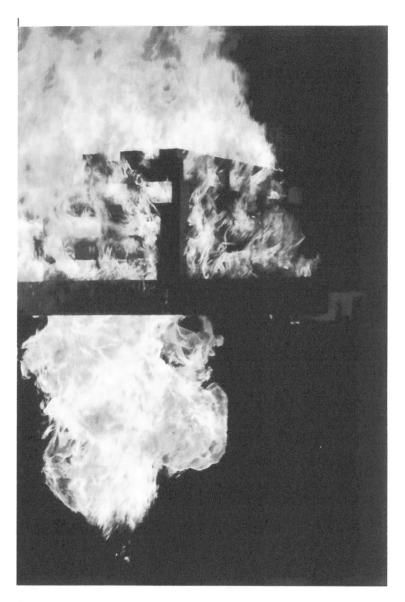

COLOR FIGURE 23.1
Fire roars through an empty wooden crate stacked within a storage arrangement.

COLOR PLATE 1.1

Firemen on duty waiting for the next alarm. As they are well aware, the peak months for home fire deaths are December, January, and February.

COLOR PLATE 1.2

A 1997 HME-Luverne rig responding to a call. It contains a 1500-gpm pump to handle challenging fires. U.S. fire departments respond to over 1.6 million fires annually. (Photo by Mike Charnota.)

COLOR PLATE 1.3

Older fire pumps came equipped with a large metal nameplate identifying the manufacturer as well as pertinent pump capacity data.

COLOR PLATE 1.4

A fire hydrant installation in progress.

COLOR PLATE 1.5
When painting sprinkler piping or equipment, it is imperative that the fire sprinklers do not receive any paint or paint spray.

COLOR PLATE 1.6
"I can think of no more stirring symbol of man's humanity to man than a fire engine." Kurt Vonnegut, Jr.

COLOR PLATE 1.7

A (supply side) O, S, & Y control valve, complete with a tamper switch and accompanying signage.

COLOR PLATE 1.8
The valve arrangement for this fire sprinkler system supply has had a backflow prevention device retrofitted in to comply with state plumbing code requirements.

COLOR PLATE 1.9
Fire does not hibernate in the winter months. Responding to this smoky fire is a 1975 American LaFrance fire engine. Like most, it carries plenty of equipment including hose wrenches, nozzles, ladders, suction apparatus, fire axes, industrial flashlights, air packs, roof saws, a generator, a thermal imager, and lots of hose. (Credit: photo by Mike Charnota.)

COLOR PLATE 1.10
Signage has been provided inside this valve room to identify the 2″ main drain valve, the O, S, & Y (outside stem and yoke) valves, and a placard has been affixed to the riser which displays pertinent hydraulic design information for the system.

COLOR PLATE 1.11
It is imperative that fire protection engineers inspect all new system installations for approval and acceptance. Their job does not end there, because an improperly maintained system diminishes in reliability over time.

COLOR PLATE 1.12

This outside entry door has been marked to identify that it houses both the fire sprinkler valves and the fire alarm control panel.

COLOR PLATE 1.13

Alarms have alerted firefighting personnel to this overnight fire. The rig shown is a 1982 Ford/Emergency One fire engine. Roughly half of all fire fatalities result from fires that occur between 10:00 p.m. and 6:00 a.m. (Credit: photo by Mike Charnota.)

COLOR PLATE 1.14
In larger cities and suburbs, a minimum of three firefighters, including one that is a paramedic, ride a rig like this one on every call.

COLOR PLATE 1.15
Volunteer fire companies have been part of Americana since 1736. The two pictured here are part of a long list of notable U.S. citizens who nobly served as volunteer fire fighters, including Thomas Jefferson, Paul Revere, Millard Fillmore, John Hancock, and James Buchanan.

COLOR PLATE 1.16

During Fire Prevention Week activities, fire departments often target fourth grade classrooms for educational talks because fourth graders are old enough to comprehend advanced concepts of fire prevention and will share what they learn with other children and also their parents.

COLOR PLATE 1.17
A deluge system riser including a system control valve and deluge valve.

COLOR PLATE 1.18
The system control valves for this bank of risers consist of grooved-end butterfly valves that indicate an open or closed position.

COLOR PLATE 1.19
A soffit being constructed to conceal fire sprinkler piping for a retrofit installation inside a church lobby.

11

Fire Pumps

Fire pumps are installed around the globe to protect buildings and property from fire loss by pumping water to standpipe systems or automatic fire sprinkler systems. These pumps make large amounts of water available for automatic fire protection use. By pumping water at boosted pressures, they ensure that fire protection systems are adequately supplied for effective operation should the need arise. These systems of protection, of which fire pumps are often an integral component, automatically detect fire, sound alarms, start the automatic firefighting operation, and remain in operation for as long as necessary to control, suppress, or extinguish the blaze.

Fire pumps are particular to those structures in which the existing local water supply pressure needs to be augmented in order to accommodate demand factors such as building height, size, construction type, and hazardous degree of contents. A fire protection engineer must determine whether a fire pump is needed and, if so, what type, where, and how it will be installed to supplement the available water supply.

If the existing water supply is of adequate volume, the pump may be necessary just to make up for a pressure deficiency. All pumps raise the existing water supply pressure, which greatly assists the sprinkler system design technician as he goes about his task of making system pipe size determinations. In all cases, a fire pump installed along with a reliable power source and an abundant water supply is desirable due to the hydraulic advantages of a readily available high-pressure water supply. The approved installation of the complete fire pump package ultimately secures lower insurance premiums for the building owner.

Five men representing various insurance companies organized the National Fire Protection Association (NFPA) Committee on Fire Pumps in 1899. The early fire pumps were manually started. What has been long documented is the fact that the possibility of fire exists in all buildings. Without the undertaking of safe fire protection measures, all structures are at some degree of risk.

Fortunately for today's pump manufacturers, fire protection is an accepted commercial building norm in the United States and, as a result, the U.S. market for fire pumps continues to flourish. The rest of the world lags behind North America in fire protection acceptance. In Europe, people will routinely pay higher commercial insurance premiums in order to avoid installing fire protection: the imprudent thinking there being that fire sprinkler systems are something that gives the buyer no return on the money. Only in recent years, and primarily in Great Britain, have overseas building authorities begun to require more stringent fire protection measures for new construction. These changes have been gradual, but each code change has increased the global fire pump demand by solid and definite increments.

Fire pumps are often referred to as booster pumps because they boost the water pressure within a piping system. High-rise buildings are an example of a structure type that typically contains a fire pump due to the inherent need there for increased water pressure. Pumps will supply two systems, sprinklers and standpipes, that normally share the stairwell risers in high-rise buildings. But contrary to the general view, these are not the only structures that are in need of fire pumps. The chief factor in ascertaining the need for a fire pump is usually the degree of the hydraulic demand of the fire protection system. Changes in the way that fire sprinklers are now manufactured have affected fire protection system design considerably over the last 20 years. Specialized types of sprinklers have been introduced successfully in the marketplace over that period of time. Certain sprinkler models have carved their own niche in the market because they have been designed to perform more effectively under specific conditions.

The ESFR (early-suppression, fast response) sprinkler is one example of this. Subsections 8.4.6 and 8.12 of NFPA #13 are concerned exclusively with requirements governing this type of fire sprinkler, which is a popular choice of protection when the building occupancy consists of high-piled storage of products. High-piling equals high danger, exponentially more dangerous by the foot. Fire within such a warehouse develops very fast and produces accelerated heat-release rates. The ESFR sprinkler head will fuse more quickly and deliver a heavy water discharge in order to suppress a potentially aggressive fire raging below. The larger orifice and deflector are features of the sprinkler that account for the broad spray pattern of water that is dense enough to penetrate and battle the fire.

However, a typical water supply will likely not be enough to satisfy the pressure and volume requirements for a system in which ESFR sprinklers are installed. ESFR and similar new types of "storage sprinklers" provide a high water density, but require very high end-head pressures. Therefore, fire pumps are required to augment the systems. Increased water supply demands often mean that a fire pump package will have to be called out and specified in the preliminary stages of construction planning. See Figure 11.1.

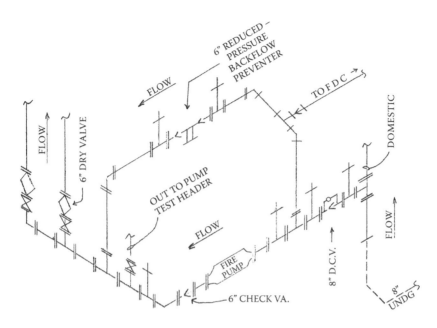

FIGURE 11.1
An isometric detail illustrating a fire pump and bypass installation.

The Horizontal Electric Fire Pump

The most common fire pump in use is the electric-driven, horizontal shaft, single-stage, centrifugal pump. This pump type is readily arranged for automatic operation. The horizontal split case pump is noted for its generous water passages, efficient operation, and easy access to all working parts.

Water flow enters at one or both sides of a bronze disc (called the impeller) from the suction inlet in the pump's casing. Power from an electric motor is directly transmitted to the pump through the shaft, which rotates the impeller at very high speeds. The larger the impeller diameter is, the greater is the rate of speed. This rotation drives the water by centrifugal force to the discharge outlet. It is the action of the centrifugal force that provides the added pressure. (See Figure 11.2).

While centrifugal pumps remain the most common pump used for fire protection purposes, other types, such as positive displacement pumps, are also available and now addressed by the standard. In fact, Chapter 5 of NFPA 20 consists entirely of new information specifically addressing positive displacement pumps, which operate by discharging a set volume of water

FIGURE 11.2
A horizontal split-case electric fire pump. (See color insert following p. 52.)

during each pump shaft revolution through a mechanical means, such as a piston plunger or rotary gear. Positive displacement pumps are most commonly used to pump foam concentrate into a foam system and for water mist systems that require very high pressures.[1]

In most cases, fire pumps must be directly coupled to the electric motor (or diesel engine) driver. Fire authorities around the globe do not allow anything other than a direct-drive unit. Only vertical turbine pumps are allowed to be driven through a speed-increasing gearbox.

Vertical-Turbine Pumps

When vertical turbine fire pumps were first designed, the whole idea was to create a device that would pump water from drilled wells, operating with suction lift. But because many reliable wells are often an inadequately small supply for standard fire pumps, the typical suction supplies for these pumps today are manmade reservoirs, wet pits, ponds, or underground tanks.

The principle of pump operation for the vertical turbine is very similar to that of the horizontal electric. However, the suction supply for the vertical turbine comes directly to the pump through a large vertical "column pipe" that features a basket suction strainer at the base. A nice feature of this pump type is that it is able to operate automatically without priming. One popular application of the vertical turbine pump today is within intermediary floors of high-rise structures, as they require a minimum of floor space. The water supply comes from break tanks that are replenished from city supplies by a fill line equipped with float-operated valves.

In-Line and End-Suction Pumps

The vertical turbine is not to be confused with the "vertical in-line" pump, which is a variant of the horizontal end-suction fire pump. Both types are space-savers, which make them ideal for retrofit applications. In addition, they are easy to maintain.

Both pumps are single-suction pumps in which the (flow) suction comes in from the end. NFPA #20 (*Standard for the Installation of Stationary Pumps for Fire Protection*) defines the end suction pump as one that has its suction nozzle on the opposite side of the casing from the stuffing box, and the face of the suction nozzle is perpendicular to the longitudinal axis of the shaft (see Figure 11.3). The "stuffing box" refers to the housing where the packing is "stuffed in."

Fluid flow makes a 90° radial turn through a radial vane impeller. It should be noted that for these vertical right-angle gear drive pumps, the Factory Mutual Research Corporation will not approve mechanical seal fire pumps, because the seals can lock up when used infrequently. Requirements for all fire pump installations can be referenced in *FM Global Property Loss Prevention Data Sheet* 3-7. The vertical in-line, shown in Figure 11.4, has the motor placed

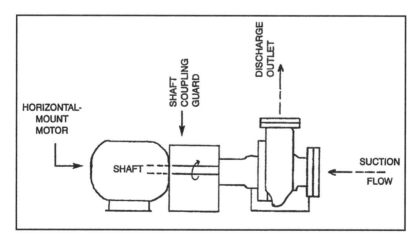

FIGURE 11.3
End-suction pump.

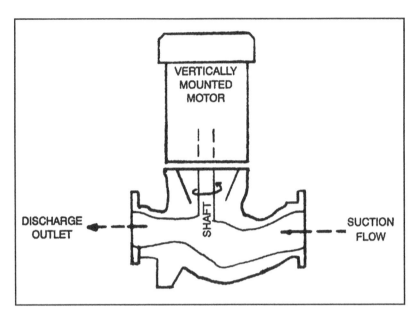

FIGURE 11.4
Vertical in-line pump.

on the pump itself. The axial motor shaft is perpendicular to the inlet. So, suction is from one end, with inlet and outlet flanges on the same linear plane; therefore the principle of centrifugal force still applies.

The difference between the two pumps is basically the motor positioning in the mounting configuration. Pump and impeller are aligned identically in each case. The vertical in-line fire pump is desirable for pump capacities of up to 750 gallons per minute (gpm) due to its compactness, and moderate price. It requires no foundation and is very reliable. The pump motor, mounted vertically, will "float" on the in-line arrangement as it expands and contracts during operation.

Diesel-Drive Pumps

Diesel-engine-driven fire pumps are a common choice in areas subject to frequent power outages. These pumps should be located in a separate pump house or pump room that is entirely cut off from the main structure. This room must be properly ventilated due to heat radiated from both the engine and exhaust piping.

Approved diesel engines for fire pump service have their rated horsepower given for a particular elevation (above sea level) with a corresponding ambient temperature. If the pump is to be installed at a higher elevation or expected temperature, the engine's useful power will be diminished. Therefore, the pump manufacturer or distributor must be aware of these conditions in order to allow for this reduction in useful power when selecting an engine. Other factors affecting diesel-drive pump equipment selection include fuel supply, the most reliable type of control, and the optimum starting and running operation of the diesel engine. Each engine must be provided with two storage battery units, each of which must possess twice the capacity needed to maintain the manufacturer's recommended cranking speed (at 40°F) throughout a 3-min "attempt to start" cycle. Battery cables are to be sized expressly in accordance with the diesel manufacturer's data sheets.

Lead-acid batteries are still acceptable and widely used. Code outlines their specific recharging procedures and current output rates. Only nickel–cadmium batteries are to be used for diesel engine drive pump projects in Europe, a much more costly installation. But the (LPC) code writers there are cognizant of the fact that facility and mechanical maintenance in Europe is typically poor. Lackadaisical facility and mechanical maintenance is also not uncommon in the United States.

Pump Selection

Once the decision is made to purchase a fire pump, the buyer's first step is to contact several fire pump distributors in the local area. The buyer will have to be cognizant of several facts that the distributor will need to know. Of primary importance: what are the minimum volume (gpm) and pressure ratings for the desired pump? Other factors include: (1) the size of the (proposed) incoming water service, (2) the suction pressure available from the existing water source, (3) the desired voltage of the pump motor, and (4) the desired hand, or impeller direction. This last item concerns pump location within the proposed piping configuration. As an example, if you are sitting on the pump motor (facing the pump), and the suction is to your left, then you will need a left-hand pump. The pump shown in the floor plan (Figure 11.5) is a left-hand pump.

A fire pump is a heavy multicomponent piece of equipment. Although installed during construction, the pump may not be put into practical use for 20 or 30 years (or more). The manufacturer is responsible for providing shop-tested fire pumps that meet NFPA conformance standards as well as those established by Factory Mutual (FM). To ensure that the pumps provide maximum reliability, with adequate pressure and volume discharge characteristics, testing agency laboratories such as Underwriters Laboratories and Factory Mutual will have investigated the properties of these products by witnessing manufacturer's tests and reviewing the data. These agencies

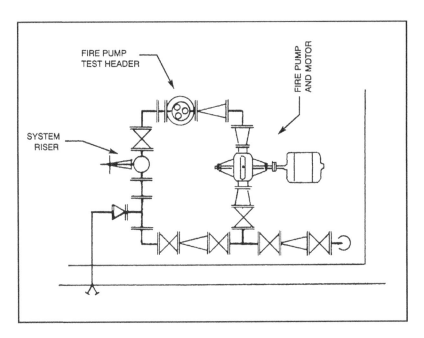

FIGURE 11.5
Fire pump and bypass.

have tested fire pumps for decades, ironing out many potential mechanical problems through their testing processes. These necessary steps are undertaken and carefully studied prior to the listing and acceptance of the various fire pumps that are on the market. Still, problems such as cracked housings, snapped shafts, or seized bearings can arise during shipment or installation. New fire pumps shipped to a job-site with incorrect pressures must be returned to the factory for modification and a new label.

Fire pumps come equipped with relief valves in order to minimize the possibility of pump damage due to overpressurization. Circulation relief valves are used to prevent a fire pump from overheating when the pump is in operation but no water is being discharged through the system. This is known as the churn condition. When operating in the churn condition, the temperature of the water in the fire pump continues to increase until it reaches the boiling point, causing significant pump damage. A circulation relief valve acts to discharge a small amount of water from the pump under churn conditions to prevent heat buildup. NFPA 20 has been revised to require that the circulation relief valve be installed on the discharge side of the pump before the discharge check valve.[2]

Proper maintenance of the fire pump and related equipment is a primary objective. Key items to check at recurring intervals include controller alarms and lights, battery cables (for diesel drives), oil or coolant levels, correct functioning of relief valves, packing gland adjustment, and checking to make sure that the motor pump and bearings are sufficiently lubricated. Of peak importance, the safety engineer must make certain that the fire pump is started and tested at regular intervals. An annual test is generally recognized as the minimum fire pump test requirement. The biggest threat to pump integrity occurs when the unit has been offline for an extended period of time. Inspectors often discover that pumps cannot start because they have "locked up." Proper maintenance and "exercising" the unit at planned intervals are the only ways to avoid this problem.

NFPA #25 recommends that electric-motor-driven pumps be run for a minimum of 10 minutes every week (30 minutes per week for diesel-engine-driven units). It also states that a preventive maintenance program should be established for all components of the fire pump assembly. Recommended testing, inspection, and maintenance parameters for pumps are also outlined in NFPA #20. The NFSA (National Fire Sprinkler Association) also has a booklet for start-up performance requirements, acceptance testing, and operation methods.

Energy for Start-up and Operation

The fire pump controller is a listed device that is completely wired, assembled, and tested prior to shipment. It must be situated within sight of the pump motor, but not so close as to "borrow trouble" by inviting the danger of escaping water from the pump relief valves, fitting connections, or system drains. Among the controller's many required features are the motor starter, external disconnect switch, a circuit breaker, alarm relay (through an independent power source), and a pressure switch that is manually set to cut in and out at pressure settings determined in the field prior to acceptance testing.

NFPA-conforming controllers must have a manual start, which must be able to operate without any other device (necessarily) actuating. For most installations, controllers are wired for manual shutdown as well. In every case, the dependability of the wiring system and electrical supply must be carefully considered to ensure a safe amperage draw. An electrical contractor must verify the entire arrangement to be in proper working order at the time of start-up.

Fire pump controllers that comprise a complete fire pump control system are constructed for use in both commercial and industrial applications, as well as for high-rise residential buildings.

The full voltage or across-the-line controller is the most economical of the starting methods. This type controller and starting method is the most simple in design and is the most widely used. When the controller receives a call to start, it engages the starting contractor for full voltage starting and a full design torque. The disadvantage to this type starting is the high demand placed on the electrical service. The electrical service shall be adequate

enough to withstand approximately 600% of full load current without causing the voltage to drop below 85% of rated voltage.[3]

The 50/60 Hertz operation relates to motor running speed. This is key,
because faster running speeds mean smaller pump equipment packages. And,
of course, pump equipment size directly relates to product price. Electrical
power supplies worldwide are roughly 33% 60 Hz and 67% 50 Hz. The 50-cycle
supply is the norm in many countries. The most popular speeds employed (but
not all) are 1,450 and 2,900 rpm for 50 Hz, and 1,750 and 3,500 rpm for 60 Hz. If
a high-power supply is needed for very high-speed pumps, yet is unavailable,
the building owner should consider a diesel-drive pump installation (if permissible by local EPA authorities). Under the NFPA codes, the electrical power
must be sufficient to cover the total predicted flow condition.

Relevant Codes

Fire pumps are manufactured in accordance with the requirements of U.L.
and F.M., and in compliance with NFPA #20. European pumps of similar
capacities are much less expensive to purchase, because of the looser requirements of the Loss Prevention Council (LPC) as compared to those of NFPA.
The LPC is an outgrowth of the FOC, a group of insurance companies that
issued fire protection rules and regulations that was disbanded in 1993 so
that a larger single incorporated standard could be established.

Butterfly valves, historically prohibited by NFPA #20 from being installed
in fire pump suction piping (they can cause excessive noise and turbulence),
are now acceptable provided they are installed 50 ft or more upstream of the
pump suction flange. By code, the relief valve must be situated between the
pump and the pump discharge check valve, and attached with its own shut-
off so that it can be replaced without disturbing pump operation.

Jockey pumps are defined in the NFPA #20 standard as pressure maintenance pumps. They have a rated (gpm) capacity not less than any normal
leakage rate, and sufficient discharge pressure to maintain a pressure boost
that is 5 to 10 psi higher than that of the main fire pump, so that the jockey
pump will start before the fire pump. This small pump is necessary to prevent the main fire pump from starting intermittently, which would over-
work that pump. It provides makeup water for incidental leakage such as
seepage at joints and fittings, or packing on valves, to replenish the interior
system. The old method of sizing the jockey pump was an easy call: 1% of the
fire pump's rating. It is advisable, though, to select a jockey pump that will
provide a supply equivalent to the flow of one sprinkler head typical of the
sprinkler system being supplied.

Section 5.14.6.1, concerned with installation, specifies that the suction piping for the pump be laid out below or at the plane of pump intake so that no

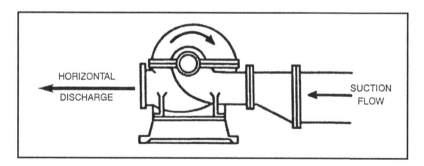

FIGURE 11.6
Horizontal split-case pump.

air pockets will form. These will cause hydrodynamic cavitation, a condition whereby air that repeatedly enters the pump casing area will eventually cause mechanical operative damage to the pump and motor. (This is a simplified explanation of what is really a very complex phenomenon, but suffice it to say that whatever the causal intricacies, the resulting condition of vapor pockets in liquid flowing through the impeller that collapses with a water hammer effect, is known as cavitation.) Horizontal centrifugal pumps are particularly susceptible to cavitation.

Cavitation can cause excessive turbulence which may lead to loosened connections and fittings. If the cavitation is severe and goes undetected, the inevitable result will be pump failure. In some situations, an eccentric tapered reducer (shown in Figure 11.6) can be installed to prevent any air pockets developing in the supply piping.

NFPA #20 warns against placing any device in the suction piping that may restrict the fire pump discharge flow, leading directly to cavitation, because pockets of vapor will form in the volute casing when decreased pressure there falls below the pressure corresponding to liquid temperature. NFPA authorities are wisely encouraging the placement of pressure-restricting devices after the pump discharge rather than before it, to avoid this occurrence. Section 5.15.6 reads as follows, "[A] listed check valve or backflow preventer shall be installed in the fire pump discharge assembly."

However, the section prohibiting any device or assembly that will restrict the fire pump's discharge flow (leading directly to cavitation) includes the following under "Exception No. 1" which reads: "Backflow prevention devices and assemblies shall be permitted where required by ... the authority having jurisdiction." Under these circumstances, final pump performance must not experience a negative-suction-pressure-read at 150% of rated capacity discharge. Obviously, when a pressure-restricting device is thrown into the whole mix, the pump's factory-certified performance test data curve (a hydraulic performance curve) is going to vary when compared to the field performance curve plotted at start-up.

Code Interpretations

Per NFPA #20, fire pumps should be housed in separate, heated, non-combustible, sprinklered rooms or small appurtenant structures. Proper ventilation and floor drainage are required. The code mandates a minimum clearance of one inch around pipe that passes through this room's walls or ceiling. Building code officials must ensure that the fire pump, driver, and controller are completely protected from adverse conditions and mechanical damage.

The fire pump room should be located as close as possible to areas where fire protection is most imperative. About 10 years ago, the NFPA #20 code committee added a new section for clarification, stating that indoor fire pump installations, including power supply and control circuits, be separated from all other areas of the building by two-hour rated fire construction. The Technical Committee on Fire Pumps voted 23 to 2 in favor of this change.

This proposal strikes a chord, inasmuch as field inspectors often report fire pumps in interior locations with little or no separation. This usually does not occur, however, when an insurance company is involved in facility planning. In addition, insurance companies often require outside access to any room within the structure that serves as the fire pump room. For example, Wausau Insurance Company's interpretive guide summarizes their own additions and exceptions to NFPA #20. Specifically, under 1-2.1 (Purpose): "All fire pump installations shall be designed by an individual NICET certified Level III or IV in Water-Based Fire Protection Systems Layout." Under 2-1.2 (Water Sources), the guide states: "Limit suction pressure in the underground supply main to a minimum 20 psi unless local purveyor allows less."

NFPA #20 notes that the discharging pressure of jockey pumps shall be "sufficient to maintain the desired fire protection system pressure." Wausau's official interpretation of this reads, "Jockey pumps shall maintain fire pump churn pressure plus any static supply pressure." These notations indicate that the insurance firm considers NFPA #20 a minimum standard. For their purposes, the word "should" in NFPA #20 always means "shall," referring to a mandatory requirement as opposed to a recommended advisory.

Other HPR carriers (most notably FM Global, IRI, CIGNA, Travelers, and Kemper) also have engineering guidelines that include amendments relative to the requirements of NFPA #20. Many require that a bypass be provided for all fire pumps, and specify a minimum distance of 10 times the pipe diameter between the pump suction flange and any (downstream) directional change fitting. They also disallow installation of closed-loop flow meter test arrangements, which are permissible under Section 5.19.2 of NFPA #20. The central reason for this is that the closed-loop flow meters have a tendency (especially over time and for larger-capacity pumps) to provide inaccurate flow-test readings.

Field Engineering Concerns

Fire pump systems started by pressure drops are generally systematized by setting the fire pump start point manually at 5–10 psi less than the jockey pump start point setting. The first settings made in the field, however, are the jockey stop and the fire pump stop points, which should approximately equal the fire pump shut-off pressure plus the minimum static suction pressure. The jockey pump (cut-in) start point is then normally set 10 psi less than the (cut-out) stop point.

It should be determined initially that the pump is primed and the casing is full of water. After starting the pump, bearings should be scrutinized for signs of overheating. Proper alignment must be observed, and foundation bolts should be checked for tightness. Any "knocking" heard from the pump, or fluctuating reads on pressure gauges, may indicate problems with the suction intake.

Size and Safety

The selection and determination of fire pump capacity is a task performed by fire protection design engineers and is primarily based on their empirical knowledge and experience. The engineer must take into account the combined input of both the local authority and the insurance company engineer. For a given demand, the pump size can be selected by dividing the total suppression system demand (including the hose allotment) by 1.5, and then selecting the next largest available pump size. This is the typical cost-effective method undertaken. Per NFPA #13, where tanks and pumps constitute a sole acceptable water source, requirements for inside and outside hose need not be considered in determining the size of the pump or tanks.

Listed pump capacities will be any of the following gpm capacities: 25, 50, 75, 100, 150, 250, 300, 400, 450, 500, 750, 1,000, 1,250, 1,500, 2,000, 2,500, 3,000, 4,000, or 5,000. Returning again to the Wausau Interpretive Guide (2-1.2), "Select the fire pump based on the total required fire protection systems demand. In no case shall the total demand exceed 120% of the rated fire pump capacity. Total demand will always include ceiling sprinklers, in-rack sprinklers and inside hose and may also include outside hose." Most insurance carriers have similar requirements. In fact, some carriers require the total demand not to exceed 100% or 110% maximum. This is more stringent than what appears in NFPA #20, where … pushing the pump all the way to 150 percent of its rated capacity in order to meet the system demand is allowed (20:5.6.6.1 and 5.8.1), it is encouraged that one does not exceed 140 percent of the rated capacity. As discussed in NFPA 20:A.5.8, when operating beyond 140 percent, the performance can be adversely affected by suction conditions. It also suggests not basing the design on less than 90 percent of the rated capacity.[4]

It is the responsibility of the design engineer to plot the combined public water supply and pump curves against the system demand. For proper pump sizing to be verified, he has two conditions to check for: (1) that the churn point (0% rated flow) combined with the suction static psi does not have a total pressure in excess of the working pressure of the system. And (2) that the 150% rated capacity design point does not draw the suction pressure to a (psi) point low enough that full pump operation may actually run the risk of collapsing city water mains. If either of these conditions presents itself, an alternative pump design is usually available.

The need to combat fires aggressively is essential to maintain life and property, and to ensure the continuation of business enterprise. In light of increased property values and expanding businesses, complete fire suppression systems augmented by properly designed fire pump packages are needed to achieve total security throughout the industrial environment. The potential for large fires poses the primary danger for which all fire protection professionals must be concerned.[5]

Endnotes

1. Milosh Puchovsky, "Peak pump performance," *NFPA Journal*, May 2000, p. 109.
2. Milosh Puchovsky, "Peak pump performance," *NFPA Journal*, May 2000, p. 110.
3. Bill M. Harvey, "Fire pumps: Problems & solutions," *Sprinkler Age*, March 2000, p. 27.
4. Roland Huggins, "Sizing fire pumps in a high-rise," *Sprinkler Age*, December 2009, p. 12.
5. Mark Bromann, "Fire Pumps: Overview for the Year 2000," *Professional Safety*, May 1999, pp. 38–41.

12

Standpipe Systems

Somewhere in the continuum that stretches between no fixed fire protection at all—and a completely sprinklered building—exists a solid point representative of the level of protection provided by standpipe systems. These fixed piping systems are found within hotels, high-rise structures, schools, jails, parking garages, and various industrial occupancies. The primary reference used when designing standpipes is National Fire Protection Association (NFPA) pamphlet #14, which defines the standpipe system as an arrangement of piping, valves, and hose connections that are located in such a manner that water can be discharged in streams or spray through attached hose nozzles for the purpose of extinguishing a fire. Somebody, usually a firefighter, opens a standpipe valve when water is needed to combat fire. In short, standpipes provide a means of manually applying water to fires within structures.

Initial efforts undertaken to develop NFPA #14 began in 1912, after the need was recognized for a reliable water source inside large buildings where fire sprinklers were not provided, and areas existed that could not be reached by hose lines attached to outside hydrants. Buildings that have very large floor areas, as well as multistory structures, are examples of facilities where the presence of standpipe systems is vital to firefighting operations.

Classifications

A Class I standpipe is one that provides a 2-1/2-in. fire hose valve on each building floor for the exclusive use of fire department personnel. As opposed to fire extinguishers (which may also be present inside a standpipe hose cabinet), the water discharge coming from standpipe hose nozzles is of almost an unlimited duration. Class II standpipe systems are accompanied by (only) 1-1/2-in. outlets, with 1-1/2-in. valves and hose, which are installed in place primarily for the use of building occupants. (The term "building occupants" was replaced by "trained personnel" in the 2003 edition of NFPA #14.) For any level of a building to be considered properly covered under Class II, NFPA #14 requires that all areas of the floor are within a 130-ft travel distance (hose path) from the hose connection proximity.

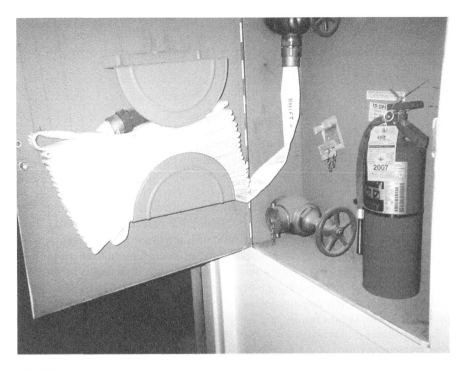

FIGURE 12.1
A Class III standpipe supplies this fire hose cabinet which houses both a 1-1/2-in. valve with lined fire hose and a 2-1/2-in. fire hose valve for use by the fire department, with room left over for a fire extinguisher. (See color insert following p. 52.)

A standpipe installed with both 1-1/2-in. and 2-1/2-in. fire hose valves is classified Class III. Firefighters may make use of the 1-1/2-in. hose lines under certain conditions. For example, this hose line becomes a valuable resource in situations when a fire occurs close to the standpipe location instead of say, the building entrance point. In such an instance, utilizing this smaller hose averts the time delay necessitated by fire service personnel having to lay hose lines hundreds of feet through the structure, prior to being able to apply water to the fire. See Figure 12.1.

System Types

By far the most common type of standpipe is the wet system, which is charged at all times with water under pressure. For unheated buildings, an automatic dry-pipe standpipe system is often utilized. Similar to a dry-pipe fire sprinkler system, water from the (connected) supply source will not enter the standpipe until the stored air pressure downstream of the dry pipe valve dips below the water supply pressure. This will occur relatively soon after one of the hose valves is opened.

If there is no permanent water supply servicing the structure (occasionally the case with stand-alone parking garages), a manual dry system can be installed. This consists of a standpipe containing air at atmospheric pressure conjoined with a permanently mounted fire department connection. In the event of a fire, the system is to be supplied with water from an arriving fire department pumper. The piping designs of manual systems are based on the water supply provided by the fire department (at the fire department connection). Standard operating procedures dictate that the responding fire service will charge the system with 150 psi at the fire department connection.

Design Essentials

NFPA #14 defines a high-rise building as one that exceeds 75 ft in height, measured from the lowest level of fire department vehicle access to the floor of the highest habitable story. The standard also specifies that manual standpipe systems shall not be used for high-rise buildings, and shall not be used in any case for Class II or III systems. Code also requires that piping for Class I and III systems must be a minimum of 4 in. in size. NFPA #14 contains a pipe schedule table for sizing standpipes, although this method can only be utilized if the system type is wet and the building is less than 75 ft in height. Generally speaking, Class I and III standpipes are hydraulically designed to provide the required water flow rate (a minimum of 500 gpm) at a minimum 100 psi at the most remote 2-1/2-in. hose valve outlet. Class II systems are only required to flow 100 gpm. The building engineer must be cognizant of the requirements of all applicable building codes, which differ from those of NFPA #14 and may include a host of tradeoffs. It's not wise to tinker with the codes, because very often more than one will apply. If the authority having jurisdiction allows for pressures less than 100 psi, particularly in fully sprinklered properties, the minimum pressure requirement is often reduced to 65 psi.

Class I and III systems comprised of just one standpipe are designed for a 500 gpm flow. An additional 250 gpm flow must be figured in for each additional standpipe, up to a total flow that need not exceed 1,250 gpm. The fire pump is to be sized accordingly. If the total floor area exceeds 80,000 sq ft, the second most remote standpipe must also accommodate 500 gpm. In instances where the standpipe also services automatic sprinklers, a separate sprinkler system demand is not required unless the water supply requirements for the sprinkler system exceed the overall standpipe system demand. The "combination sprinkler/standpipe" is a natural for fire protection, because design parameters of NFPA #14 mandate a reliable source of water for fire sprinklers to be placed on each floor of a building. This has proven to be of considerable economic (and design) benefit to building owners contemplating retrofit sprinkler projects. What simplifies hydraulic calculation procedures for fire sprinkler system designers in these instances is the knowledge that the calculation need only be run to the existing standpipe source, where 500 gpm at 65 psi is a given on any floor.

The 2006 edition of the International Building Code (F 905.3.1) requires that a Class III standpipe system be installed throughout buildings where (the floor level of) the highest story is located more than 30 ft vertically from the lowest expected (grade level) access of fire department vehicles. One of the exceptions to this requirement is that the standpipe may be Class I when the building is completely equipped with fire sprinklers.

Suppose you are looking to add fire sprinklers in the basement level of an 18-story building equipped with a fire pump and (one) standpipe. Because the static read on the uppermost floor of the high-rise (if constructed after 1990) is most likely 100 psi, then the static psi at the basement standpipe connection surely exceeds 175 psi. This is probably more pressure than you desire (and in excess of the working pressure of ordinary pipe fittings). You could tap into dedicated fire protection piping upstream of the fire pump for your sprinkler supply, thereby supplying the new sprinklers with city water only. However, you would likely run into problems with arranging the fire department connections in doing so, which would only confuse system design and compromise safety as a consequence. The most viable solution in this case would be to tie in to the existing standpipe using a pressure reducing valve (PRV). This component automatically prevents excessive pressure without dumping water, and regardless of changing flow rates or inlet water pressure surges. A control circuit is adjusted by the tension of a pilot spring to a lowered pressure aggregate. While in service (at 0 gpm flow) the pilot control system (on most PRV models) opens or closes to maintain the desired downstream pressure setting. Pressures created from the pump during churn should always be considered when determining if PRV(s) are needed.

The latest (2010) edition of NFPA #14 requires the installation of an approved device to regulate or reduce both static and residual standpipe pressures where hose connections available for occupant use have static pressures in excess of 175 psi. No more than 200 psi of water pressure shall be held at any connection, to further safeguard firefighters from using hose lines under overpressurized conditions. Underwriters Laboratory requires the installation of pressure gauges on both sides of a PRV, and also a 1/2-in. relief valve on the downstream side. The valve is to be inspected, tested, and maintained (per NFPA #25) just like any other system component. It is recommended that the PRV be inspected quarterly for leaks, thread damage, downstream pressure maintenance, and to validate that it is in an open position.

Periodically inspecting all portions of standpipe systems is an important, indispensable responsibility. The following is a brief overview of other standpipe system inspection requirements:

Weekly

Ensure that all control valves are in the open position.

Quarterly

Inspect all hose rack assemblies; note age and type.
Check condition of all hose and nozzles.
Ensure that all hose couplings turn freely.
Ensure that all caps are in place.

Semiannually

Test all system tamper switches.

Annually

Check hose cabinet overall condition.
Conduct a main drain test.
Conduct a thorough inspection of all exposed pipe.
Inspect and test all valves.

Five-Year Interval

Conduct a standpipe flow test (one riser at a time).
Hydrostatically test for leakage on dry manual standpipes.
Test Class II and III occupant hose.

Always bear in mind that a reliable water supply is the most essential element of any standpipe system, so it must be well maintained and re-examined at regular intervals. Because standpipe system characteristics may be complex, and varying codes may apply, the input of the local authority is an absolute necessity when the basic system is on the drawing board. The standpipe is a lifeline to safety, and whether it follows that building integrity and guarantees the long-term security of its occupants depends on its proper maintenance and upkeep.[1]

Endnote

1. Mark Bromann, "Usage of Common Standpipes," *PM Engineer,* April 2004, pp. 21, 22, 24.

13

Antifreeze Systems for Unheated Areas

Including those slated in residential occupancies, over 90% of all sprinkler systems installed today are wet-pipe systems. But not all building interiors are climate-controlled. Rising energy costs have led many to consider leaving portions of their buildings unheated. When this occurs, the only fire protection system options available are dry-pipe, preaction, heat tracing, or antifreeze. The traditional antifreeze system has been defined as "A wet-pipe system that contains an antifreeze solution and is connected to a larger wet-pipe system that it uses for a water supply. It supplies automatic sprinklers to protect small unheated areas."[1] Where the unheated area is large, it is generally understood that the most economically viable alternative is to augment the wet-pipe system with an auxiliary dry-pipe system to protect the area subject to freezing temperatures.

The design of an antifreeze system depends on whether it includes a backflow prevention device. If it doesn't, the arrangement of the supply piping and the valves is shown in the Figure 13.1 cross-section. In this detail, designated points #3 and #4 refer to a 3/4-in. gate valve and fill cup, #6 refers to a supervised indicating valve, #8 refers to a 1/2-in. test valve, #5 denotes a 1-in. auxiliary drain valve, and #10 refers to a check valve with a 1/32-in. hole drilled in the clapper. Often, an installer will simply shave off a corner of the clapper instead, to allow for fluid expansion. National Fire Protection Association (NFPA) #13 allows the check valve to be omitted entirely if all antifreeze piping falls below the elevation of the indicating valve (all antifreeze solutions are heavier than water). The two test valves (noted as #8) must be installed at least 48 in. apart. It should be noted that the gravity-fed fill cup may be placed almost anywhere downstream of the check valve and is normally installed at a high point of the piping. (See Figure 13.3.)

Figure 13.2 is the NFPA #13 required supply arrangement for antifreeze systems containing a backflow preventer. In most jurisdictions, this device must be a reduced-pressure type. This component will prohibit water from flowing back into the water supply, but overpressurization can occur when pressure builds in the system as a result of thermal expansion caused by temperature change. When the fluid expands, the excess is discharged into the expansion chamber. Within the vessel, the gas-charged bladder compresses as the fluid enters the shell. All system components are thereby protected because the chamber allows the trapped antifreeze to expand. Expansion

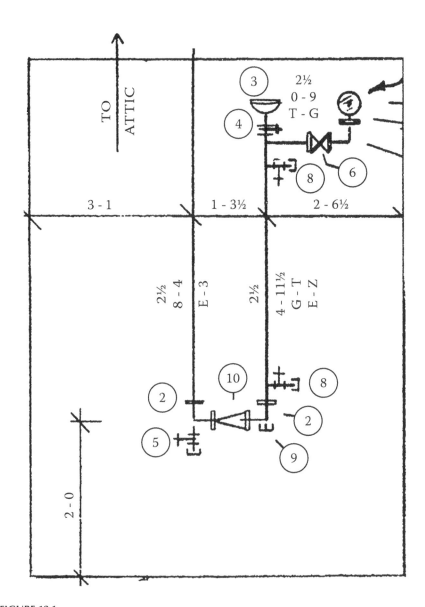

FIGURE 13.1
The scaled shop drawing of an antifreeze loop depicts, in part, two 1/2" welded outlets for 1/2" test valves (designated by the number 8), which were delivered to the project on a 2-1/2" thread/groove piece of steel piping.

tanks vary in size depending on the volume of solution contained in the anti-freeze system, and they must be listed and marked accordingly. They must also be sized to handle system capacity. (See also Figure 13.4.)

The two types of antifreeze are propylene glycol and glycerin. Section 7.6.2.4 of NFPA #13 states that "An antifreeze solution shall be prepared with a

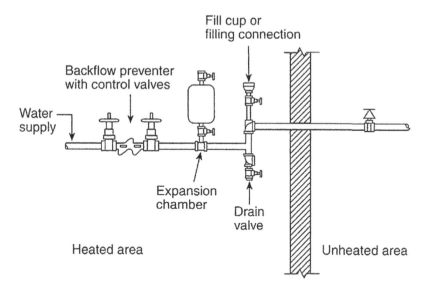

FIGURE 13.2
A two-line sectional diagram of an antifreeze system valve arrangement. The expansion chamber is pre charged with air pressure.

freezing point below the expected minimum temperature for the locality." As an example, the freezing point for the Chicago area is –22°F, which means that the glycerin solution used there would contain 60% glycerin and 40% water. If propylene glycol (PG) is used, the correct mixture would be 50/50. The solution must be thoroughly mixed with water before filling any system. An over-concentration of antifreeze would be costly and may sabotage the protection desired, because beyond a certain point (around 75%) of overconcentration the antifreeze will thicken and cease to lower the freezing point of the solution. Installers should know (this is common sense) not to mix any solutions in containers or drums that may be contaminated with other substances.

It is imperative that the system be tagged with the manufacturer of the antifreeze, including all pertinent details, nuances, and specifications of the fluid. When the solution is field-tested (annually, as required by NFPA #25), testers can determine the applicable concentration through the use of (it is hoped) reliable laboratory-grade hydrometers or refractometers of sufficient quality. For hydrometers to produce accurate readings, the fluid temperature must be between 60 and 68°F, or else a temperature correction should be applied to the reading. After five years or so following installation, the solution may break down due to small leaks and concentrate will have to be added. After a stagnation period that comprises eight to ten years without servicing, the system will require complete solution replacement. If not properly maintained, the far ends of the piping will no longer have the suitable mixture. When any section of the system contains inadequate antifreeze, freezing and breakage of sprinklers, pipe, and fittings will result. Not only will this comprise a filthy and expensive mess, a frozen system cannot suppress a fire. See Figure 13.3.

If the volume of an antifreeze system exceeds 40 gal it is required that an additional drain/test connection be added at the remote system proximity, to provide another point for solution concentrate level testing. For systems with a capacity exceeding 150 gal (see NFPA 13: Section 7.6.3.7), an additional test valve must be added for each 100-gal increment. This requirement addresses the reality that the solution mix is most likely to break down at the far reaches of the system.

In spite of the fact that antifreeze systems have been comprised of steel or copper piping for years, they are generally not installed using CPVC piping. When they are, only glycerin can be used (per NFPA #13 and #25 requirements) as PG solutions will chemically attack CPVC pipe joining compounds. With glycerin solutions, contamination may arise from the use of recycled or off-grade glycerin. Also, only thread paste sealant compatible with the particular CPVC piping material is allowed. Sealants and gaskets used in CPVC piped systems must also be checked for compatibility, and most manufacturers provide chemical compatibility websites such as http://www.systemcompatible.com to resolve any questions posed by the consulting engineer. Petroleum or solvent-based lubricants or sealants must never be implemented. Mixing steel and CPVC in the same system is generally a bad idea. For systems of steel or CPVC, field survey reports have found that rubber-gasketed adjustable drop nipples installed in antifreeze loops are prone to leakage.

Engineers make things fit into the budget. Prior to the 2002 edition of NFPA #13, the standard included the advisory, "Because of the cost of refilling the system or replenishing to compensate for small leaks, it is advisable to use small dry valves where more than 40 gal are to be supplied." That section has disappeared from the code, meaning that it's now permissible to supply 50 sprinklers or more with an antifreeze system. But because solution costs exceed $20 per gallon, why would you do that? Over time, a dry-pipe system will require less maintenance, fewer service calls, and is the safe, "greener" system with regard to potable water. But if the system in an unheated area exceeds 30 sprinklers, it may still be advisable to choose an antifreeze system for any of the following reasons:

- To avoid the purchase of an excessive number of dry pendent sprinklers
- If the new piping arrangement will contain numerous low points, to avoid the installation of a plethora of drum-drip assemblies
- If the noise from an air compressor (i.e., behind a theatrical stage) is not acceptable to the owner

FIGURE 13.3
Concentration levels of antifreeze fluid are tested in this traditional "loop" arrangement (see Figure 13.1) through the two 1/2-in. gate valves.

FIGURE 13.4
This arrangement (see Figure 13.2) includes an expansion chamber. The fill cup and valve are located at the system high point in an adjacent unheated parking garage.

- Where there is virtually no floor space available for a dry valve
- Where plenum space is so tight that the potential dry pipe cannot be properly pitched
- Where any additional electrical wiring has not been planned for or desired

One other advantage of the antifreeze system is that, as the years progress, the interior of the piping will not scale as it would on a dry system. The corresponding reduction in corrosion then, will extend the life of the piping system. Also, the antifreeze system has a life safety advantage considering that, with dry systems, there is a longer lag time (up to 60 seconds) for water to hit a spreading fire after a sprinkler fuses.[2] And, there is a small increase in the probability for system failure with the dry system setup. With antifreeze systems, all that has to work is the sprinkler itself. Barring any accidental valve closures, the system will work efficiently to quickly extinguish any outbreak of fire.

Endnotes

1. Mark Bromann, "The Design and Layout of Fire Sprinkler Systems," *Technomic Publishing Company, Inc.*, 2001, p. 283.
2. Mark Bromann, "The Facts About Anti-Freeze Systems," *PM Engineer,* January 2009, pp. 9–12.

14

The Lifespan of a Sprinkler System

Let us begin with corrosion. Corrosion is best understood as a force of nature to be reckoned with. Not only is the external corrosion that slowly works on buried ductile iron water supply piping a tangible risk, but the internal pipe and fittings of an automatic fire sprinkler system must also struggle to withstand the ravages of time. The plain fact known to the trained mechanical engineer is that water is corrosive to both steel (pipe) and cast iron (fittings). This is primarily why there are recommended minimum dimensions for wall thickness of piping. The thinness of the wall of the piping used is in inverse proportion to the lifespan of that pipe.

Although this chapter is focused on fire sprinkler systems comprised of steel pipe (because those are by far the most widely used) it should be noted that copper is more resistant to corrosion than steel, cast iron, or zinc. Systems employing copper tube will enjoy a long life. Type "K" (marked with a green stripe) has a thicker wall than Type "M" or "L" copper tube. On another note, it should be observed that any systems that combine both CPVC and steel piping may become problematic. Chemicals contained within certain products used in conjunction with steel piping installations, such as thread sealants and cutting oils, have been known to lead to CPVC piping damage and degradation over time.

The Threat of Corrosion

After sprinkler system water lays stagnant in piping for several years, there is very little difference in the rate of corrosion between steel and cast iron. As facility managers and maintenance engineers are well aware, the first leaks to occur in any sprinkler system originate at the threaded ends of the piping (adjacent to fittings), where the pipe wall is the thinnest. We know that the life expectancy of the average toilet is 50 years. That is a fairly long shelf life, considering that the next toilet I buy will most likely outlive me. But a newly hired building engineer may well be responsible for the upkeep, inspection, testing, and maintenance of a sprinkler system that was installed 70 or 80 years ago. Certainly he or she must be on the alert for signs that corrosion is beginning to rear its ugly head. Corrosion naturally destroys metal at a very slow rate. When water (being conductive) comes into contact with metal, electrochemical reactions take place between the metal and the water. Rust (iron oxide) is formed as a direct result of these reactions. It may "pit" the inside pipe wall, or build up accumulated corrosion by-product deposits (tubercles) there, which inhibit optimum water flow. If not removed, this condition of blocking or "choking" the required water flow will worsen (sprinkler heads themselves can also become internally corroded or clogged, as can fittings). If undetected, corrosion will begin to overtake a sprinkler system to the point where it may become necessary for certain sections of piping to be removed and replaced. And a system in the advanced stages of corrosion will keep corroding, lessening the expected life of the newly replaced sections.

Some preventive medicine in the form of early corrosion discovery is the building engineer's key strategy. You can be alerted to difficulties if one or more of the following peculiarities occurs:

- Heavy discoloration of discharged water
- Plugged drops, drain lines, or service connections
- Recurrent pinhole leaks
- Bad sulfur smells
- Reduced flows during test procedures
- Deposits, scale, or nodules in discharged water
- "Weeping" at pipe joints

Piping Particulars

The most important element in an equation designed to increase longevity of a sprinkler system is the use of pristine potable water. If dirty, or if its quality is otherwise in question, the water used for sprinkler systems may have to be treated. The second most critical element pertains to the wall thickness of the piping used. Schedule 40 steel pipe is certainly preferable to Schedule 10, dyna-thread, SuperFlo, XL, dyna-flow, or any of the other thin-wall pipe options available in the marketplace today, especially if they are to be installed with threaded ends. The system is only as strong as its weakest link. In other words, the thinnest pipe wall is often the first to go. However, detailed field examinations have consistently reported that piping near the top of a (wet-pipe) system, or that which is closer to the water source, is also at high risk depending on the type of corrosion. (On a wet system, pockets of air will rise to the high areas of the system. On a dry-pipe system, piping near the low areas is at a higher corrosive risk, because that is where water will accumulate.) Dry sprinkler systems are associated with some of the most advanced piping failures ever documented. Due to inadequate grade there is always some moisture and water within the pipe: in reality it is never 100% "dry," and in many cases the result is slow but gradual pipe corrosion.

The life expectancy of wet-pipe sprinkler systems exceeds that of dry-pipe systems because they are wet, and thereby not subject to incremental interior pipe wall scaling that acts like a cancer in dry systems. Of course, wet-pipe systems contain very little oxygen, which reacts with water and steel to create a buildup of rust (ferrous oxides). This oxygen cell-type corrosion is extremely common in sprinkler systems. For many years, engineers expressed surprise that the pipe and fittings of sprinkler systems lasted as long as they did without incident. More recently, as some Schedule 40 systems developed leaks after just five years, it was recognized that too much oxygen had been introduced into these systems. The blame fell on either (a) numerous instances in which the system was refilled, or (b) excessive testing. Any testing over and above National Fire Protection Association (NFPA) #25 recommended intervals or insurance carrier guidelines will most assuredly give cause for the added abundance of oxygen to rust system piping prematurely.

M.I.C.

Pipe wall perforations may also be caused by microbiologically induced corrosion (MIC), a relatively rare occurrence in sprinkler systems, in which waterborne bacteria in the system water "attack" the steel (or other metals, but not CPVC). In cases where one or more of these (there are at least 11) MIC bacteria types have entered sprinkler systems, the introduction of the contaminant is often traced all the way back to the construction phase. However, bacteria known to cause corrosion in steel can also be found in lakes or in wells containing water of inferior quality. The "infected" MIC systems contain micro-organisms that feed on nutrients in the water, grow in number, contribute corrosion, and accelerate the development of the corrosion process. MIC develops in colonies, which is why the pipe degradation is localized within the system. All sprinkler systems contain some level of micro-organisms, therefore it's a little trickier to discern specifically if a corrosion condition is definitely MIC. Even if a test sample of system water reveals that many microbes and nutrients are present, you still cannot conclude that a MIC problem exists. If it does, MIC has the potential to destroy fire sprinkler piping at a corrosion rate of .10 in. of steel per year.

Periodic inspections for MIC combine a couple of standard tests so that testing agency laboratories can make an accurate assessment. Water test specimens are taken for analysis from several areas of the system (before the system is drained). Second, an internal inspection of a section of system piping is carried out. Accumulated deposits attached to the pipe can be scanned, checked, and evaluated for metallurgical properties. Orange or black tubercles may red-flag the MIC condition. The entire system is then thoroughly flushed to remove sludge, dirt, oil, pipe dope, grease, scale, and the like. If the system is MIC-contaminated, it is possible to treat the system chemically, although that step may be cost-prohibitive.

The MIC problem has been known for 35 years or so, however, it had not been detected inside an automatic fire sprinkler system until the mid-1980s. Power plants and other utilities have employed full-time corrosion engineers for decades, and they are well versed in the intricacies of detecting and treating MIC in various mechanical piping systems. Although it can occur virtually anywhere, incidents of MIC tend to be more prevalent in areas with warmer climates. Several organizations today continue to research, monitor, and study the problem diligently. Again, hard data confirming known causes of microbiological corrosion in sprinkler systems is relatively uncommon. Despite the recent wave of attention given to MIC, the sky is not falling.

On-Site Quality Control

Concern regarding the possibility of corrosion is much more common when a sprinkler system is installed in an area where a corrosive atmosphere exists. For these instances NFPA #13 requires the installation of corrosion-resistant sprinkler heads. Stainless steel sprinklers are the most durable of the available types, and by far the most expensive. Other fire sprinkler coatings available include wax, lead, wax over lead, and lead for the frame with wax protecting the fusible elements. These are listed in temperature ratings ranging from 135 to 500°F (wax melts prior to sprinkler activation; lead melts when heated to 620°F). For the exterior of sprinkler pipe, shop paint or manufacturer's antimicrobial coatings will help some, but it has been documented that almost all overhead system piping failures corrode from the inside of the pipe (pipe coatings might really be there for "show"). Also, any film of rust (oxide layer) that is sometimes seen on pipe and fitting exteriors actually serves as an excellent corrosion inhibitor. Nonetheless, the general rule for piping installed outdoors or substantially exposed to atmospheric conditions, is that it should be galvanized, painted, or otherwise protected as dictated by the particular environment.

Many factors can come into play when surmising the purported length of time that sprinkler system pipe and fittings can safely last. You could even make a case that system pipe installed by a robust fitter, who makes up each pipe length an extra turn, will extend system life a little longer just because he puts more threads into each fitting. The type of black steel (domestic or imported) used to manufacture the pipe matters very little, because all commercial quality carbon steel rusts at the same rate. (Steel containing certain alloys that slow corrosion is currently manufactured, but is not used by fire sprinkler installation firms. Only by going to galvanized steel can you realistically enhance system longevity through choice of material used.) Rather, it's mainly what goes into the pipe (clean water!) that will best determine a sprinkler system's life expectancy. For years I have been entertained by stories from installers and inspectors, describing items of debris found in systems (typically lodged in a check valve) that solved a blockage mystery. These individuals have removed nuts and bolts, two-by-fours, rocks, and canvas bags from inside some of the piping in existing systems. So I suppose it comes as no surprise when others find tiny microbes and nodules under the microscope.

Sprinkler heads are virtually guaranteed for 50 years of service. Steel is inherently strong and it doesn't age. You can say the same for cast iron fittings. Their only real enemy is corrosion. That is the natural phenomenon to be understood, monitored, and dealt with tenaciously. Limit corrosion effectively, and you've got yourself a fire sprinkler system of high caliber that will stay in service for a long time.[1]

Endnote

1. Mark Bromann, "Forecasting the Lifespan of a Sprinkler System," *PM Engineer,* June 2003, pp. 20, 22, 24.

15

Fire Protection for Computer and Electrical Rooms

A question commonly proposed to fire protection engineers called on to design a fire protection system for a new facility goes like this, "What type of fire protection is appropriate for our computer rooms and data storage areas?" Yesterday's answer was halon. The answer to the question depends on many factors and in fact, leads to practical and viable answers given today that may very well be quite different from the answer given five or ten years from now. Today, there are agencies worldwide testing and evaluating new alternative extinguishing agents to halons with due respect to concerns about environmental safety, ozone depletion, and global warming potential. The key elemental determinants being tested with regard to these inert gas agents include their level of fire suppression effectiveness, their effect on the environment, their potential for overpressurization of an enclosure, whether they produce gaseous by-products, and how quickly they activate after the origin of a fire.

Computer Rooms

It wasn't too long ago that computer rooms around the globe represented an immense market for Halon 1301 systems, widely recognized as the most effective chemical fire suppression agent developed. But today, halogenized agents are not a protection option inasmuch as it is now known that all halons are (to a small extent) ozone depleters. The bromine atom in Halon 1301 reacts readily to destroy ozone molecules. Certain halocarbon agents that do not contain the bromine atom are considered by some to be short-term "band-aid" substitutes because they have global warming potential, and will be further federally regulated. The production of halon itself essentially ended as of January 1, 1994.

Another nonoption for protection in the computer room itself is carbon dioxide extinguishing systems. Because CO_2 displaces oxygen, its discharge could cause asphyxiation or suffocation to any occupants present. A low-pressure carbon dioxide system, which would cost about double what a halon system would to install, could be installed only in the computer room subfloor space. This is environmentally acceptable, and a discharge would require no cleanup whatsoever.

With the phase-out of Halon 1301 (and 1211), several chemical companies set out to find suitable alternate replacement agents. No one has ever really come up with a "drop-in" pound-for-pound mimic of halon. The first qualification to be realized is U.S. Environmental Protection Agency (EPA) approval, particularly from their Atmospheric Protection Division. The Clean Air Act requires the U.S. EPA to research and report to the public the status of potential halon substitutes. The U.S. EPA discloses their findings in the Significant New Alternatives Policy (SNAP) publication. The EPA designates agents as acceptable, unacceptable, or pending. Along with this, the National Fire Protection Association (NFPA) has established a new technical committee. In 1991 the committee's task was to prepare a design standard to be ready for the commercialization of the agents. Now known as NFPA #2001, *Standard on Clean-Agent Extinguishing Systems,* this internationally accepted standard shows which agents are acceptable for occupied space protection. This determination is made by comparing the minimum design concentration for the agent with the highest acceptable toxicity exposure. NFPA 2001 requires that inert gas systems be designed with an agent concentration of 43% or less and an oxygen concentration of 12% or more if personnel are not able to evacuate the area within one minute.

Currently, NFPA #2001 (and SNAP) has recognized 11 halon replacement agents. Two that are in wide use today are IG-541 (better known as Inergen) and HFC-227 (commonly called FM-200). Both are used for the protection of sensitive electronic equipment and other highly critical environments not suitable for water-based extinguishment. Each agent has its pros and cons with respect to chemical composition, equipment hardware, installation restrictions, and system recharge cost, just to mention a few of the characteristics to be considered.

From the onset, the most pertinent question is, what value do you place on the operation of the computer room? Does your business depend on uninterrupted 24-hour operation of the computer room in order to survive? Typically, the suppression system desired must be comprised, in part, of the following essential features:

- An agent that is nontoxic
- A fast-responding system
- A system that does not bring about a major cleanup problem
- An electrically nonconductive medium

In computer rooms, the loss potential to take into account includes not only the destruction of computer hardware, but the loss of computer downtime. If continued operation of the computer room is imperative, then you will want to investigate the (fairly costly) installation of quick-responding chemical agent systems. Once more, two of the more popular gases that are both habitable (people-safe) and environmentally friendly are FM200 (produced by Great Lakes Chemical) and Inergen (made by Ansul). If either system does activate, the resultant down-time will be very short.

Inergen is a very clean agent, which makes it a popular choice for the subfloor space, for switchgear rooms, vaults, tape storage, process equipment, and any electronic area containing irreplaceable equipment. It is also widely used in computer rooms. It is an inert gas with superior flow characteristics, and its discharge time is anywhere from 45 seconds to a full minute. It is a mixture of three naturally occurring atmospheric gases: (52%) nitrogen, (40%) argon, and (8%) CO_2. The Inergen gas curtails and extinguishes fire by lowering the oxygen content beneath the level that supports combustion. But it should be noted that due to the CO_2 present in Inergen, the brain continues to receive the same amount of oxygen in an Inergen atmosphere as it would in a normal atmosphere for a reasonable period of time. Also, the agent does not produce a fog, so visibility within a compartment remains adequate for evacuation purposes.

Designed for versatility, Inergen cylinder valves can be opened electrically, pneumatically, or manually. Inergen has zero ozone depletion potential (ODP), zero global warming potential (GWP), and a zero atmospheric lifetime. When Inergen is released, its components simply resume their normal role in the earth's life cycle. Its installation requires the presence of one or more large alloy steel cylinders (similar to that of CO_2 cylinders), which hold the gas at vapor pressures in excess of 2,000 psi at 70°F. The amount of agent required is calculated in cubic feet rather than pounds. Pressure reducers in the cylinder manifold are a necessary system component, as is the use of Schedule 80 steel pipe up to the union orifice where pressure is reduced. Because Inergen is required to reach 95% of its design concentration within 45 seconds, and due to its higher vapor pressure, the Inergen fire suppressant cylinders can be stored a fairly long distance from the protected hazard utilizing smaller diameter discharge piping. The major advantage of an Inergen system is its quick response and the larger extent to which it can be reliably effective.

FM200 is a chemical blend (heptafluoropropane), stored as a liquid within the agent cylinder similar to that of halon-type cylinders. FM200 has zero ODP, a GWP of 0.3 to 0.6, and an atmospheric lifetime of 31–42 years. This is in strict compliance with environmental regulations. It is thermally stable and on the SNAP list. The FM-200 discharge piping utilizes Schedule 40 pipe. FM200 requires less of a footprint to hold agent in a cylinder. The supply tank will be proportionately smaller. It is held in pipe at approximately 60 psi, and discharges quickly for a duration of about 10 sec to achieve a 7% concentration. Due to the typical 7 to 10-1/2% concentration design, the agent storage reserve requires minimal space. But due to the relatively low (59–60 psi) vapor pressure, FM-200 cylinders are required to be within close proximity of the hazard. Actually, the physical properties of FM200 considered along with its traditional extinguishing requirements, allow its use with the same type of equipment that would be used for Halon 1301, requiring minimal hardware alteration for retrofit situations. Very often the same detection and control panels can be used.

Installation of the first commercial FM-200 system began in December of 1992. In addition to computer rooms, typical applications for this clean gaseous agent include telecommunication equipment facilities, data processing libraries, emergency power facilities, flammable liquids storage rooms, museums, clean rooms, process control centers, and so on. It does create hydrogen chloride as a by-product, which may attack existing documents or other "archive" materials. It will not, however, corrode sensitive electronic equipment, and contains no particulates or oily residues. In fact, it leaves very little residue and is a quite popular extinguishing agent in use today for the protection of computer rooms. Proponents of FM-200 boast code-compliant, cost-effective system installations.

Working with tight budget considerations, the potential installation of a preaction system in the computer areas should be investigated. BOCA (Building Officials and Code Administrators) defines a preaction system as a fire sprinkler system employing automatic sprinklers attached to a piping system containing air with supplemental fire detection that is installed in the same area as the sprinklers. Actuation of the fire detection or smoke detection system automatically opens a valve that permits water to flow into the sprinkler piping system and to be discharged from any open sprinklers (see Chapter 6). Preaction systems historically have had limited use in electronics-intensive applications, where the mere mention of the word "water" is probably dangerous. Some modifications are advisable. As an example, a time delay can be built into the preaction control panel, creating a time lapse before water commences to fill the system. This delay provides the building staff with a chance to investigate the incident prior to the introduction of water at the scene. As additional safety measures, on/off sprinkler heads and floor drains may be installed. Also, smoke detectors can be cross-zoned, so that two detectors must go into alarm before the preaction system is charged with water.

The less expensive option of a preaction system is really insurance for protection of the building structure itself. The loss of the entire building will be prevented effectively by a properly installed preaction system, although the degree of water damage may sacrifice the equipment data center in the process. The loss of down-time may be considerable. However, the preaction system will not activate until there surely is fire present. The question presented to the business owner is always, "What type of protection do you want?"

Again, the type of protection generally desired for a computer room is total flooding fire extinguishment, suppression that occurs quickly before any damaging smoke is generated, and without wetting equipment and contents of the enclosed room. The desired qualities of gaseous agents present near the fire origin must include quick-response time, near-zero ozone depletion, short atmospheric life, and low toxicity. If fire suppression is accomplished within 10 seconds of system activation, the release of toxic or corrosive gases (acidic products of decomposition under fire conditions) will be minimized.

When you need to guard absolutely against water damage to sensitive electronic gear or high-value areas, a gaseous agent is a wise choice for fire protection. The question of which agent is best suited is not a simple one, neither is the "better" gas, nor are any of the 11 recognized by NFPA #2001, the most optimum for all possible applications. Many issues with respect to hazard definition, size of hazard, agent cylinder storage size, locations, installation cost, and agent recharge cost must be addressed. As an example, although the FM-200 gas may cost 8–10% less up front, it has a higher replacement cost. When choosing between FM200 and Inergen, you have to look at the total situation, talk to a professional, and make a choice.

The effectiveness of clean-agent systems is always dependent on the integrity of system interaction with process controls. In addition, an adequate amount of agent to supply must be determined, in accordance with recommended design concentrations for the cubic-foot volume of the computer room. With regard to clean-agent system design, there are several red flags that must be avoided:

- Room smoke detectors must not be placed within close proximity of air diffusers.
- Doors must be self-closing, and never blocked "open."
- Inadequate pressure venting may result in overpressurization of the enclosure.
- Too much air-handling equipment may result in loss of agent during discharge.
- Cable and ventilation openings must be minimized.
- Incomplete fire separations will allow exposure from external fires.
- Concentration levels (especially for flammable-liquids hazards) must be adequate enough to combat skin fires.
- Deficient fire barriers at the ceiling will fail to contain the gaseous agent.

One final option for computer room protection currently available is water mist systems, discussed in Chapter 6. The efficiency of these systems is well documented, and they may be economically advantageous as halon alternatives in many applications. Their fine water spray absorbs heat, cools flame by diluting oxygen with steam, and reduces overall heat intensity. The water-cooling effect is enhanced as a result of the division of water into small drops, which also maximizes evaporation. The creation of fine droplets increases the surface area available for heat absorption. Similar to the result firefighters obtain with fog streams, the mist allows a fuller interaction with air currents, which will scatter the droplets over a larger area, blocking the transfer of radiant heat to nearby combustibles. This process will extinguish fire without causing unacceptable levels of water damage.

A National Fire Protection Association technical committee on Water-Mist Fire Protection Systems (NFPA #750) is continually developing performance criteria to evaluate adaptability of mist systems to various specific applications. Strainers are used liberally within these systems to guard against potential nozzle clogging problems, because of the small orifices used in the spray nozzles. The piping system must also possess a very high corrosion resistance for the same reason. Water-mist systems are a beneficial fire protection tool for applications where neither gaseous agents nor standard conventional sprinkler systems are a satisfactory answer. The evolution of fine spray technology has shown significant progress in successful fire testing, making this another viable alternative.[1] Water-mist system design and installation requirements can be found in the *FM Global Property Loss Prevention Data Sheets* 4-2.

New solutions take time: consider the fact that it took the industry 20 years to get halons developed and out into the marketplace. An aggressive research and development program is being conducted by the U.S. Department of Defense. Their goal is to demonstrate environmentally friendly and user-safe processes, techniques, fluids, and agents that meet their own operational requirements (once satisfied by halon systems) for their aircraft, ships, combat vehicles, and critical-mission support facilities. National Fire Protection Association Standard #75, *Standard for Protection of Electronic/Data Processing Equipment*, outlines the basic regulations governing computer room protection. An informed choice for the selection of the most appropriate fire detection and suppression system for computer rooms is a critical step for commercial buildings. Regardless of the system selected, containment of the protected space and separating it from the rest of the building is equally important.

Electrical Rooms

In terms of code intent and interpretation, you'll have to scratch your head to find a room that has comprised a "grayer" area over the years than that of the electrical room in a commercial building. If the building is sprinklered, should the electrical room contain a fire sprinkler or not? This question has been posed to authorities having jurisdiction around the country for decades, resulting in diametrically opposite answers for at least as long. The City of Chicago Fire Prevention Bureau, as one example, flip-flopped on this issue so many times that fire sprinkler contractors routinely designed their sprinkler systems without a fire sprinkler shown in the electrical closet, but with 1-in. piping extended to just beyond the room's exterior, terminating in a 1-in. threaded cap, just in case the Bureau changed their tune at the last minute and demanded the inclusion of a sprinkler.

At the seat of the ongoing controversy is the old maxim that water and electricity don't mix. Echoing this reality, insurance companies have long disallowed the presence of any fire sprinkler piping inside electrical control rooms, citing "good safety and fire prevention practices." The only time this has not been a tough call for AHJs (Authorities Having Jurisdiction) it seems, is when the building in question did not have a fire sprinkler system. Tiring of endless requests for information (RFIs), many jurisdictions eventually put their intent down in writing. Section 15-16-350 of the Chicago Building Code now reads that

> [S]prinklers shall be provided throughout the entire premises, except in rooms used solely as electrical equipment rooms containing generators, transformers, or switchboards (unless such equipment is located in a public utility structure). Sprinklers may be omitted from rooms or areas containing materials which react violently to the application of water. ... Sprinklers shall be installed throughout telephone exchanges except in rooms housing switching, toll, main distribution frame, power, auxiliary power or switchboard equipment.

In contrast, many cities outline their requirements either online or in a published text that simply mimics the NFPA #13 verbiage, stating that "[S]prinkler protection shall be required in electrical equipment rooms," unless (Section 8.15.10.3) all of the following conditions are met: "(1) [T]he room is dedicated to electrical equipment only, (2) only dry-type electrical equipment is used, (3) equipment is installed in a 2-hour fire-rated enclosure including protection for penetrations, (4) no combustible storage is permitted to be stored in the room." And where sprinklers are installed, Section 8.15.10.2 adds that "hoods or shields installed to protect important electrical equipment from sprinkler discharge shall be non-combustible." Several firms manufacture preshaped flat tiles to cover conduits and cable trays for protection of electrical cables and equipment in the event of a fire. The core of the NFPA Technical Committee concern is that over time, generator or transformer rooms frequently become places to store combustible materials. Any accrued buildup of dust, greasy debris, or the presence of trash containers only serves to amplify the problem. The intent of NFPA #13 is to avoid the omission of sprinklers in a room just because electrical equipment is present.

It's not so much a debate as it is a problem rife with complications. Regulatory mandates notwithstanding, the generally understood safe advisory is never to put water on an electrical fire (ideally, use a multipurpose fire extinguisher). When the electrical room must be sprinklered by code, an engineering design firm will normally require a (fixed-temperature) heat detector for the room, and a manual pull station in close proximity, separately zoned at the fire alarm control panel. These passive fire protection measures serve to alert occupants of a fire at an early stage. The design firm will also call for the sprinkler(s) to be 212 or 286°F temperature rated, to avoid sprinkler discharge in the event of a high (closed room) temperature increase. Because any accidental discharge of water is to be averted, head-guard protection is also provided for the sprinklers.

Electricity travels in a closed loop, or circuit, and flows easily through conductors, taking the quickest path to the ground. Water is an excellent conductor and the human body consists of mostly water, thus a body becomes electricity's path to the ground if that person contacts water that touches electricity. In addition, electricity will jump out of its intended circuit where convenient, which injures or kills thousands of people every year.

Many fires result from faults in the power delivery system itself. Causes of these defects may arise from faulty installation, degradation due to aging, physical damage, or overloading. Electrical devices also pose a fire risk due to bad switches or internal wiring failures. When fire arrives from another source, electrical equipment is very susceptible to damage from the heat and smoke (containing corrosive products of combustion) produced by that fire. Many electronic components begin to fail at approximately 174°F, with major component failure arising when temperatures exceed 300°F. The ideal scenario, of course, is when a Class "C" (or A-B-C) fire extinguisher exists in the electrical room, and is used effectively on a small fire. No one should attempt to extinguish a large fire with an extinguisher (see Chapter 10). If one is not available, the only course of action left would be to evacuate, attempt to shut down the main power source (by experienced personnel, and assuming that power supply is not uninterruptible), and when calling 911 be sure to let them know that this is an electrical fire.

There are several options concerning suppression system choice. An automatic CO_2 flooding system discharging CO_2 gas into an electrical room will stop any fire. But as previously discussed, this gas is lethal at use concentrations in confined spaces. Also, equipment damage can still result due to the toxicity and conductivity of carbon dioxide. A clean agent fire suppression system, such as Inergen, FM-200, or Sapphire would be the safer alternative. In either case, the room volume is calculated to determine exactly how much agent (or percentage of CO_2 gas) will be required for (normally) 10 minutes of total flooding. The primary purpose of a fire sprinkler (or preaction) system is to douse or confine the fire through the application of water to its area or room of origin in order to protect the structure. With a gaseous clean agent system, the primary function is to provide protection for the sensitive (and probably valuable) assets within the enclosure being protected. In addition to high system cost, the downside of clean agent systems for these applications is the stringent requirement of intact enclosures with doors shut and external ventilation completely sealed prior to agent discharge.

If the electrical or "utilities" room takes up less than 800 cubic feet, with a ceiling height not exceeding 12 feet, a single self-contained "Cease Fire" overhead fire extinguishing system unit may be installed. Formerly referred to as "halon balls," these EPA- and FM-approved cylinders are designed for single-room protection, activate automatically (using fusible links), and flood the room with a nontoxic blended powder and gas agent or Halon 1301 (the halon type which is still environmentally safe). They are rechargeable and require no piping. This type of protection is endorsed under the "Other Automatic Extinguishing Equipment" section (7-7.3) in NFPA #101.

A high-pressure water mist system may also be implemented for fire protection in electrical rooms. With small discharged water droplets, the water mist system employs much less water than a conventional sprinkler system and certainly less than would be received from firefighting operations. The ability of a water mist system to stop a fire actually increases with the size of the fire due to the extent of evaporation. Although that provides clear security, problems can arise if the fire is small. First of all, there will be a limited evaporation of water droplets in a small room. Secondly, because water mist does not provide "total flooding," system discharge cannot navigate around cabinet doors or other potential obstructions.

Applicable codes to reference for protection recommendations include NFPA #72 National Fire Alarm Code, NFPA #10 Portable Fire Extinguishers, NFPA #75 Protection of Information Technology Equipment, NFPA #2001 Clean Agent Fire Extinguishing Systems, and NFPA #70 National Electrical Code. Requirements for fire-resistive enclosures in rooms containing circuit breaker panels, transformers, and the like, as well as additional fire protection requirements for special high-voltage electrical equipment, can be found in NFPA #70. The fuel load of a typical electrical room will include electronic equipment and the conduits and wiring necessary to supply power to various electronic equipment contained within the structure. To minimize risk, no additional unnecessary fuel loading should exist in the electrical room. Prevention and suppression of any fire occurring around electrical systems is realized primarily through careful planning and the resourcefulness of the engineer. Depending on room size, equipment voltage, and the overall degree of the hazards involved, a logical system design choice must be ascertained to provide a dependable working solution to service the fire protection needs of every area containing electrical service within a facility.

Endnote

1. Mark Bromann, "Fire Protection for Computer Rooms," *PM Engineer,* June 1999, pp. 26, 28, 30, 32, 33.

16

Kitchens

Fire potential exists within any structure containing a kitchen, and that chance is greater than you may think. From 2003 to 2006, 75% of structure fires reported in dormitories, fraternities, sororities, and barracks involved cooking equipment (and were most common during evening hours between 5:00 and 11:00 p.m.). Kitchens (at 20%) are also the leading area of origin in U.S. office properties. In hotels and motels, cooking equipment is the leading cause of fires (37%). Cooking is the number one cause of fires in churches. From high schools to day care centers, the list goes on. The greatest threat occurs when kitchen cooking is unattended. Those unguarded moments are the most prone to disaster. People who have ever put out a kitchen fire with a fire extinguisher will tell you that they were right there at the onset of flames.

Commercial Kitchens

The following is taken from an investigator's report of a 1994 kitchen hood fire that escalated into a full-scale conflagration inside a commercial building.

> At the time of the loss, the cooks were preparing chicken on the gas grill under the vented hood. When they added cooking oil a fire started in the grill area and moved up into the overhead vent area. The Ansul fire control system operated as designed. However, the fire got out of the vent area and into an enclosed space above the drop ceiling. From here, the fire traveled into other concealed spaces in the building. There was inadequate water and the sprinkler system was improperly designed allowing the fire to destroy the building. The building was a total loss.

In saying "improperly designed," what the report refers to is a (vertical) 9-in. combustible concealed space that was not protected with automatic sprinklers, a definite oversight. The report continues:

> The pictures clearly show that there were holes drilled into the overhead hood, as well as the vent leading through the concealed space and to the roof. These holes were drilled to allow passage for the electrical cables used for hood lighting and the roof vent blower. These holes allowed two things to happen. First, over time, they allowed grease to escape from the hood and to coat the outside of the hood and the vent inside the concealed

space. This gave the fire a source of fuel which was easily ignited. Second, the holes allowed the fire to travel outside of the hood and the vent and into the concealed space between the roof and the drop ceiling.

This particular blaze was so intense that in order to obtain adequate water to fight the fire, the fire department ordered the sprinkler system shut down upon their arrival. Their efforts were ultimately in vain, and three firemen were injured in the process. What followed in the aftermath was a series of fire investigations and litigation involving inspectors, cleaning companies, and the installers of both in-place fire protection systems.

This study is a rare instance in which a combination of mishaps and blunders (holes in the kitchen hood, delayed 911 call, 22 psi water supply, accumulated layers of old grease, heavy dose of olive oil on the grill, etc.) led to danger and ultimately, disaster. In cases like this one, no one ever really figures out exactly what happened. An eddy current from the fan *may* have forced air vapors out of the holes. The grease *may* have ignited just from the heat and not necessarily from open flames. Duct system behavior is very unpredictable, so who knows what happened there except to say that a perfect exhaust system was breached by the holes in the hood. It's safe to assume that the concealed space was filled with vapors at the onset of ignition, meaning that the presence of automatic sprinklers in that space (above the hood) would have been critical for fire mitigation.

Results turned out infinitely better for the owners of a fully sprinklered truck stop in LaPine, Oregon. A 5 am fire there was attributed to a commercial range burner that was accidentally left on overnight under a pot of griddle grease. The ensuing range and grill fire was completely suppressed by an Ansul automatic suppression system in the hood. No fire sprinklers activated in this blaze, and the hood exhaust system had even blown all smoke out of the restaurant by the time the first fire engine arrived. Fire Marshal Jim Gustafson noted that the kitchen hood suppression system "kept the fire extremely small and extinguished it." As opposed to the first example, this restaurant was unoccupied at the time of the fire. Even so, the business suffered only minimal down-time, taken up by cleaning crews and for the restoration of the hood suppression system.

Kitchen Hoods

For all commercial kitchen hoods, the pinnacle of protection technology is the wet chemical system. Over 90% of kitchen hood systems installed today utilize this effective agent. Like other manufactured options, the Ansul R-102 system is pre-engineered and utilizes a potassium-based wet chemical. Discharging from all nozzles simultaneously, the agent gives a foam "look," covering a grease fire with what appears as a foam blanket. Activation is accomplished by either (1) manual detection (with a remote hand pull), (2) automatic (with a fusible link), or (3) electrical detection. The primary system components are the control head, cylinder, and discharge nozzles. With the Ansul system, the cylinder is pressurized via a cartridge. Cylinders provided by other manufacturers, such as Range Guard and Kidde, come pressurized at around 175 psi. The systems are typically tied in to the building's fire alarm system, to signal the appropriate station upon system activation. Also activated are components that will shut down the electrical and gas fuel sources, and the hood exhaust system. It is important that the heat sensor or any alternative method of system activation shuts down air, so that the majority of discharged chemical agent is not exhausted. Also, if the air exhaust of the hood continued to operate it would draw room oxygen towards the fire, thus making matters worse for fire extinguishment.

The Piranha system, dubbed a "hybrid" system, discharges the same wet chemical agent (or similar variants), followed by water. About 5% of new installations use hybrid systems for hood protection. Quench systems that combine wet chemical foam and water act like a deluge system, and come with a special head that delivers the agent in more of a mist form. The less popular "water spray" system option discharges water only into the fire area (of the hood, duct, or plenum) and is connected to the wet-pipe sprinkler system of a building. Once upon a time, dry chemical systems were used for kitchen hoods, but these have not been manufactured since the mid-1980s, due to changes in the UL300 standards. New installations of dry chemical hood systems have not been allowed since 1994.

Water mist systems are a poor choice for kitchen hoods, because those nozzles are easily clogged with oily sludge and not easily cleaned. Most recommended nozzles manufactured specifically for kitchen hoods come with blow off caps, and run about $32 per nozzle. Cylinders are manufactured in 3.0, 4.6, and 6.0 gallon sizes (or delineated by weight).

Exhaust Systems

All restaurant cooking appliances produce heat, and this obvious fire hazard is increased when there is the possibility for grease accumulation and other cooking by-products. Exhaust systems for ovens, broilers, deep fat fryers, and the like must be equipped with a grease removal device. These consist of filters and special fans, designed to remove vapors and to provide a fire barrier. Safe and effective ventilation of fixed cooking equipment is a tricky business, and recommended design requirements are covered in National Fire Protection Association (NFPA) #96. National code requirements for kitchen-hood wet chemical systems are addressed in NFPA #17A.

Previous editions of NFPA #96 required the movement of air exhaust at a minimum velocity of 1,500 ft/min. However, recent HVAC engineering studies have demonstrated that lower velocities tend to keep airborne materials moving whereas the higher velocities actually result in a higher buildup of cooking grease. As a result, NFPA #96 now requires a maximum exhaust velocity of 500 ft/min. One result of this change is that the associated ductwork will have to be larger to handle the lower air velocity flow. Another code update mandates the certification of hood and duct cleaning contractors, a move applauded by insurance carriers. NFPA #96 also requires that additional signs and placards be displayed in languages that are in harmony with that of the appropriate restaurant staff.

Food Preparation

More cooking is accomplished today with vegetable oils as opposed to animal fats, in line with changing dietary trends. But because vegetable oils cook at higher temperatures, restaurant equipment stays hotter longer and there is an increased chance of reignition with those types of grease fires. To adapt, the industry has developed wet chemical extinguishing agents that cool the equipment as they suppress the fire, thereby minimizing any chance for reignition. In addition to the cooling effect, the agent smothers the fuel surface and separates liquid fuel from vapors.

There are several maintenance issues regarding commercial kitchens that are keys to fire safety:

1. Do not leave any appliance unsupervised.
2. Keep areas around ovens, grills, and deep fat fryers clean, free of grease, and clear of combustible objects.
3. Review cooking appliance operational and safety instructions with staff on a regular basis.
4. Check pilots of stoves to ensure that they are lit.
5. Ensure that high-temperature thermostats in deep fat fryers are functioning.
6. Ensure that exhaust system is in operation whenever cooking equipment is on.
7. Never use flammable cleaning products or solvents.
8. Post instructions for the manual operation of any hood suppression system.
9. Have all equipment professionally cleaned at frequent intervals.

Local and national codes and insurance company requirements generally dictate a 6-month interval for kitchen hood inspections. At a bare minimum, fusible links are required to be replaced every 12 months. The exhaust system should be checked weekly for greasy film deposits or residues. Inspection service agreements are typically available from providers of portable extinguishers, kitchen hood suppression system contractors, and a select few fire service companies.

People who work in restaurants aren't trained as firefighters, and the presence of in-place fire protection systems offsets this obvious fact. But a kitchen hood suppression system is only one ingredient of fire-safe operations in a commercial kitchen. The weak link in commercial kitchens with regard to fire prevention is often inspection, testing, and maintenance operations. The NFPA pamphlets require about 25 separate steps for each 6-month kitchen hood inspection. Adherence to this regimen makes for a sense of security.[1] When recommended servicing and maintenance procedures are ignored, the consequences can be severe.

Residential Kitchens

The NFPA reports a total of 3,430 (civilian) fatalities in the United States in 2007 resulting from fire. Of those, a whopping 2,895 (that's 84%) occurred in residential occupancies. When we begin to look at the residential fire problem, we must start with the fact that more fires originate in the kitchen than any other room in the house or apartment. And the leading cause of kitchen fires comes in the friendly disguise of stoves and other cooking appliances. In 2005, the Ohio State Fire Marshal's Office noted that cooking fires accounted for 25% of the year's 126,347 residential fires. The United States Fire Administration states that 32% of all home fires begin in the kitchen area. Statistics from the NFPA cite that cooking equipment, at 28%, is the leading cause of residential fires (heating equipment landed in second place with 14%). When you factor in the NFPA estimate that cooking fires account for more than half of all unreported residential fires, this equates to at least one cooking fire incident per year per eight American households. Among those households, those having gas ranges have a lower risk of fire and associated losses than those that use electric ranges.

The most significant statistic agreed upon by every interested agency is that the number one major cause of all cooking fires is their unattended operation. And the cause of that very dangerous oversight is distraction by a variety of day-to-day activities. These include household chores, a phone call, tending to a child, a television program, or some similar event that absorbs the attention of the person preparing food to such an extent that he or she may even forget that something is burning or boiling on the range. The diagnosis is that multitasking while cooking leads to human error. A recent study documented that the greatest number of domestic arguments in the average American home take place in the kitchen. On the flip side of that, fire officials constantly remind the public that the greatest threat to life safety occurs when no one is in the kitchen at all. People today lead busy lives. Data have shown that in 68% of cooking fires, the individual operating the cooking appliance is somewhere in the residence other than the kitchen.

An Accident Waiting to Happen

The ongoing and troubling reality of careless, poorly supervised, and unattended cooking fires in apartments and homes is so prodigious and widespread that it comes as no surprise that this problem has been thoroughly researched. The Association of Home Appliance Manufacturers, for example, has conducted extensive studies which reveal that the person responsible for 50% of home kitchen fires was between the ages of 30 and 49. Their findings further show that in the great majority of instances, what first ignited was either grease or oil, food left on the stovetop, or combustible materials near the cooking appliance itself. More facts have come to light in the past several years:

- People inside the residence evacuated the premises without attempting to quell the blaze in 64% of kitchen fires.
- Approximately half of the people who attempted to extinguish the fire did so incorrectly, further compounding the danger.
- The actual time of cooking fires peaks at 6:00 pm. A secondary peak occurs at noon.
- The average dollar amount of damages resulting from all cooking fires is $1,573.00 per incident.
- On Thanksgiving, the number of residential fire occurrences increases by 12% from the typical daily average, with unattended cooking being the causal factor at a rate of almost double the daily average.
- The range-top is involved in four out of five cooking fires.
- Twenty-five percent of cooking fires are caused by men or women between the ages of 19 and 30.
- Forty-one percent of people who perished in home cooking fires were asleep at the time of the fire.

- Working smoke detectors save lives and also double a person's chances of escaping a nighttime fire. In 2002, a smoke alarm was present and alerted occupants in 45% of all cooking fires.
- Fifty-seven percent of home cooking fire injuries are suffered by victims who attempt to fight the blaze themselves.
- By property type, 54% of home cooking fires occur in one- and two-family dwellings; 46% occur in multifamily or other residential occupancies.

The last item is noteworthy and consequential because an unconfined fire in an apartment or townhouse will directly affect other residents of the structure. It should be noted that in 2005 there were kitchen fires originating in areas other than the stove that resulted in 30 civilian deaths. Most often, these were traced to electrical malfunctions in refrigerators, freezers, dishwashers, or the microwave oven. Microwaves are generally safe. They cause water molecules in the food to vibrate at high speeds, creating heat to cook the food. The reason that glass or ceramic plates don't heat up in a microwave oven is because they contain no water molecules. However, if a twist-tie or other piece of metal is accidentally placed inside the oven, arcing may occur which poses a very real threat.

Keep an Eye on the Stove

A typical calamity occurred in October, 2007 in a Burlington, Vermont (eight-unit) apartment building. A resident was arranging her table settings in a separate room while heating cooking oil in a saucepan. Eventually the oil ignited and when the fire intensified this person removed the flaming pot from the stove, put it in the sink and ran water. This ill-advised action (mixing water with burning oil) resulted in a sharp explosion that scattered flames onto the walls and cabinets. Fortunately for the woman who inadvertently spread the fire, her apartment was equipped with a residential fire sprinkler system, and a single sprinkler activated, extinguishing the blaze prior to the arrival of the fire department. Others are less fortunate. A 2001 kitchen fire in South Carolina killed a 25-year-old woman and her two young children. The woman had fallen asleep with the stove on and, as is typical in kitchen fires, the flames from the stove first ignited the cabinet immediately above, quickly spreading fire throughout the single-family home that was devoid of any fire sprinklers or smoke alarms.

In the Event of a Fire

In case of an oven fire, always close the oven door and turn off the oven. If an appliance is electrical, unplug it if reasonably possible. If something is on fire inside a microwave, leave the door closed until the flames are out. If the fire is on the stovetop, smother the flames with a larger pan or a lid after protecting your hands with an oven mitt or a dish towel (don't wet a wrapped towel, scalding may result if the moisture in the towel gets too hot). Better yet, use a throw rug to smother the fire if one is lying nearby. If your clothes catch fire, stop, drop, and roll. Do not throw flour on a grease fire; use the fire extinguisher. Never move, carry, or even touch a flaming pot. Never ever attempt to transport a burning deep-fat fryer. If the fire is not immediately brought under control, evacuate the premises and call 911.[2]

On the Alert

Awareness is the key to prevention. Cooking fires are preventable and simply should not happen. Something as simple as looking for signs of overheating may avert a household disaster. Basic fire prevention tips for the home kitchen include:

- Check cooking devices for any cracks or fraying on cords and plugs.
- Unplug any small cooking appliance when not in use.
- Because it's easy to turn on the wrong burner, remove any items from the stove when not in use, especially pans of cooking fat, gravy, or oils.
- Keep pot handles turned toward the back of the stove.
- Keep the stove and oven clean of grease and food particle buildup.
- With children in the home use the back burners, particularly when boiling water. Keep pets out of the kitchen and forcefully establish a (3-ft) safety zone around the stove for children.
- Keep all oven mitts, wood utensils, grocery bags, dish towels, curtains, papers, and all other combustibles far away from cooking appliances.
- Stay in the kitchen when frying, broiling, boiling, or grilling food.
- If baking, roasting, or simmering, make regular checks and always stay close to the kitchen. Make regular use of the timer.
- Double-check that all dials on ovens are turned off when the cooking is finished.
- Never cook while intoxicated.

The fact that the consequences of a home fire can be severe cannot be over-emphasized. Two very important rules to remember are (1) that a dry chemical fire extinguisher belongs somewhere in the kitchen, and (2) that your primary responsibility when encountering any home fire is to get yourself and the other occupants out of the house as soon as possible. Don't be hesitant to share this information with your family, friends, and neighbors.

Endnotes

1. Mark Bromann, "For the Sake of Flame Broiled Food," *PM Engineer*, June 2004, pp. 38, 40, 42.
2. Mark Bromann, "Keeping the Heat Out of the Kitchen," *PM Engineer*, December 2008, pp. 18, 20, 22.

17

Lightning Protection

For businesses nationwide, a total prevention of damage from lightning strikes would save millions of dollars each year in facility and content damage, and would also serve to safeguard critical company processes. Lightning strikes are random, and well-known for their highly destructive nature. Some areas of the United States are more prone to this type of natural disaster than others. In Chicago, for instance, statistics show an occurrence of about 40 thunderstorms per year. Although thunderstorms appear less frequently (about 10 per year) in California, the likelihood is that Florida can expect anywhere from 75 to 100 thunderstorms annually. The necessity for lightning protection, however, is in direct proportion to the severity of the thunderstorms. The hazard is greater in open areas as opposed to cities with a high density of buildings, where it should be recognized that taller structures are most at risk. The threat of lightning exposure is also high for buildings with a large footprint. The danger posed by lightning peaks in the summer months, typically occurring in the late afternoon or early evening. Lightning is least likely to occur in the winter season. From 1997 through 2006, lightning deaths were most prevalent in the following U.S. states (in order): Florida, Colorado, Texas, Georgia, and North Carolina.

Lightning can start fire(s) with a single strike, and the roofs of wood-frame buildings (with combustible attic spaces) can be a spot of origin, as can (combustible) cedar wall siding. The lightning strike is often accompanied by a brilliant flash of light followed by a loud shocking clap of thunder. If the building in question contains many occupants, a panic may ensue just from the commotion. This issue arises periodically because many places of public venue are often the tallest structures, and lightning will almost always strike a point closest to the base of the thundercloud. "Sideflash" can occur, which consists of lightning jumping between two metallic objects. Lightning, a very fast traveling atmospheric discharge of electricity, is always accompanied by thunder.

The potential for fatality should not be of great concern to occupants of a building because, during a thunderstorm, inside is the safest place to be. Of those victims killed by lightning (an average of 55 annually in the United States), all are generally in outdoor locations. In 1902, a famous lightning strike damaged the upper section of the Eiffel Tower, which required reconstruction. Portions of a building most susceptible to lightning strikes include water towers, church spires, flagpoles, deck railings, chimneys, parapet walls, and skylights. If the roof is flat, the edge of the roof is where the lightning will probably strike. Fires can occur if lightning hits a flammable liquid or gas tank, producing sparks that ignite escaping tank vapors. But regardless of the specific cause, the ensuing fire will cause much more damage than the lightning strike itself. Historic buildings and landmarks that are considered to be of civic value cannot be replaced and should always be equipped with lightning protection.

An interesting side note in U.S. history is that when Benjamin Franklin said, "An ounce of prevention is worth a pound of cure," he was explaining why he had just attached a lightning rod to his house. Today, many commercial insurers require lightning protection. A system of (U.L. certified) lightning protection gives businesses a measure of safety, ensuring that the maximum protection has been afforded to their building's exterior. The systems serve to intercept lightning before it can strike a building, tower, or tank, and then discharge the current into the earth. Installers must follow National Fire Protection Association (NFPA) requirements stipulated in NFPA #780, *Standard for the Installation of Lightning Protection Systems*. This pamphlet also outlines regulations governing procedures for inspection and maintenance.

The concept of lightning protection is to ground the strike to control the electrical charge without loss or damage to property. Inasmuch as metals are good electrical conductors, a continuous metal path (isolated from other metals) must be created from an air terminal (the traditional "lightning rod" component) to the ground terminal. There are several types of lightning arresters that will prevent the flow of ordinary electric current over this path. If the metal is of sufficient size to carry the expected (lightning) current, it will be virtually unaffected by the heat and related forces it receives. A direct path is best, constructed of a nonferrous metal such as aluminum or copper. This large diameter cable is called the main conductor. Precautions must be taken to protect metals from corrosion, and nothing shall be painted. If alloys of copper are used, they must be substantially resistant to corrosion as well. The structural metal framing of the building itself can be used as part of the lightning protection system (but not a metal roof). The air terminal must be installed at an elevation at least 10 in. above the highest point of the building to avoid damage from arcing. Each main conductor is attached to a metal grounding rod that is typically set 10 ft into the earth. To prevent wall damage, each ground connection should extend a minimum of 2 ft away from the foundation wall, and 2 ft beneath it. To further protect the structure, a ground loop conductor is installed which interconnects main conductors, down conductors, and ground terminals. NFPA #780 addresses requirements for the number of air and ground terminals, and required spacing, depending on roof pitch and roof configuration. The systems are to be thoroughly inspected every five years. In particular, ground connections must be checked because rods may become corroded or broken at a level just beneath grade. Damage to the system may also result from snow, high winds, reroofing, falling tree limbs, and from various forms of building remodeling.

Only reputable experts familiar with the nationally recognized safety and installation standards should design lightning protection systems. All systems are customized to fit the needs of the particular structure being protected. They function 24/7, during all conditions including power outages. Surge protection devices and service entrance arresters can become part of the package. A surge arrester is defined in NFPA #780 as "a protective device for limiting surge voltages by discharging or bypassing surge current. It also prevents continued flow of follow current while remaining capable of repeating these functions."

For personal safety during a thunderstorm from lightning, or a lightning "splash" or sparks from a sideflash, one should seek shelter in one of the following:

- A city street surrounded or shielded by nearby buildings
- Inside a tunnel or subway
- Inside a metal-framed building
- Inside an automobile, bus, or train
- Inside the nearest large building

In any event, do not remain in any of the following locations:

- An open field, tennis court, or a golf course
- A parking lot
- An open boat
- In a lake or swimming pool
- Under an isolated tree
- Near any wire fence or railroad track

And make sure that you are not inside a convertible car, golf cart, or farm machinery. Steer clear of electrical appliances, open windows, and any metal objects. From 2002 through 2006, 22% of all fatal lightning strikes occurred to a victim situated beneath a tree. If in a rural area outdoors, your safest shelter will be a depressed crevice or valley, or else in dense woods. Drop to your knees and keep your hands placed on your thighs; do not place your hands on the ground, and do not lie on the ground. Roughly 9 out of 10 injuries and deaths caused by lightning occur in rural areas.

Underwriter's Laboratories maintains a Master Label Certification of lightning protection materials and equipment to ensure high safety levels. Inasmuch as thousands of buildings are damaged by lightning each year, this and similar efforts are implemented to minimize property loss and to protect businesses against down-time. Secondary effects from lightning, such as electromagnetic pulses, can wreak havoc with sensitive electric circuitry within a building. Voltage from a single bolt of lightning can travel within a building and ignite certain flexible gas piping (if it is present) known as corrugated stainless steel tubing. Also, lightning that strikes electrical overhead wires may be conducted to a building's interior through those wires.

Lightning is a compelling danger and it is a leading cause of fires. Lightning strikes do not necessarily follow a typical pattern. Although the NFPA guidelines do not provide a 100% guarantee against the risk of lightning damage, a brief reading of NFPA #780 will serve to illustrate just how complex the protective measures of a networked lightning protection system must be. Contractors for this specialized discipline should be certified by the (LPI) Lightning Protection Institute. Any new system must be installed by and under the supervision of qualified professionals well-versed in this technology. This is extremely vital for any structure containing highly hazardous materials.

18

Hotels and Motels

On February 13, 1986, an eight-hour inferno at the luxurious Manila Hotel in the Philippines took the lives of 27 people. Two months later (April 18, 1986) a fire swept through the upper floors of a newly built hotel in Harbin, China, resulting in 10 fatalities. Also in 1986, an afternoon fire at the Dupont Plaza hotel in San Juan, Puerto Rico claimed the lives of 97 more. For most, this is news that is just hard to stomach.

What is known from a fire safety standpoint in the United States is that it is statistically safer to stay at a hotel than in your own home. Nonetheless, the United States is no stranger to major hotel tragedy. This was evidenced by the early morning fire of November 21, 1980 at the MGM Grand in Las Vegas (85 killed, 700 injured) and the evening fire of February 10, 1981 at the Las Vegas Hilton (8 killed, 198 injured). More recently, a 54-year old man (Robert Vasquez) succumbed to fire in his second-floor room at the Mason Hotel in downtown San Diego in the early hours of December 18, 2004. Careless smoking was cited as the cause; 17 were injured. One nagging little detail: none of these hotels mentioned contained a fire sprinkler system.

The fire investigator's report on the Dupont Plaza Hotel fire (later determined to be caused by arson) noted that all 97 victims had perished within 12 minutes of the onset of the blaze. This fact alone is testament to how quickly fires in hotels can become lethal monsters, releasing toxic products of combustion capable of extinguishing life in as little as two minutes. Test reports have shown that in an average-size room, a small fire can ignite all the room's contents in under five minutes. Incidental to the lodging industry is the fact that hotels house large numbers of occupants who may be sound asleep. When awakened by loud alarms, disoriented lodgers must first figure out where they are and then frantically attempt to locate a viable escape route in a large unfamiliar, burning structure.

Let's lay out a few basics for the traveling public:

- Today, about 3,900 fires occur annually in U.S. hotels and motels, half of which are small confined fires
- Only 50% of all U.S. hotels (regardless of size) have fire sprinklers.
- There were 39 fire fatalities in U.S. high-rise motels between 1985 and 1998. There were 45 fire fatalities in all U.S. hotels and motels from 2005 through 2007.

- Hotel safety and security "scores" vary widely. Newer hotels normally score higher. Small inns and B&Bs generally score quite low. Large upscale hotels usually rate higher scores than smaller economy and midprice hotels.
- One in seven U.S. high-rise hotels still lacks automatic fire sprinklers.

The hotel industry is well aware that growing numbers of the general public are cognizant of the life safety risks posed by unsprinklered high-rise hotels. When I worked for Viking, one of our well-traveled executives adamantly refused to stay in a hotel that did not come sprinkler-equipped. A female executive with whom I've worked doesn't take it that far, but always demands a room on either the second or third floor in a high-rise hotel, purely in the interests of life safety. And, the 1990 *Hotel and Motel Fire Safety Act* mandates that all federal employees stay only in hotels that are National Fire Protection Association (NFPA) #13- and #72-compliant for smoke detectors and automatic fire sprinkler systems. Anyone wishing to play it safe today can reference the FEMA Hotel-Motel Master List on http://www.usa.fema.gov/hotel to determine if a hotel meets their standards.

In February, 2004, on the heels of a five-story Comfort Inn fire that killed six, State Senator Verne Smith introduced a bill that would require all hotels in his home state (South Carolina) to have fire sprinklers installed by 2007. Smith contended publicly that the bill was badly needed, but it was defeated the following April. Senator Greer remarked that he was unable to overcome opposition from a group of Charleston lawmakers who felt the bill would "be onerous" for the historic hotels of their city. As a concession, what became law was a requirement that all nonsprinklered hotels post a (8-1/2 × 11) sign near the registration desk for their guests, stating that their facility is not required to have sprinklers. Senator David Thomas said that requirement would shift the burden and responsibility of fire protection to consumers, who could choose for themselves.

Moving westward, the city of San Francisco unanimously passed a measure in July of 2001 that gave owners of residential hotels one year to install fire sprinklers throughout their facilities. Residential hotel fires in the previous four years had claimed three lives and destroyed 11 hotels. The bill's chief sponsor, Gavin Newsom, observed that "All of these hotels had one thing in common: lack of sprinklers," adding, "this legislation is long overdue." The bill affected residential hotels over two stories high and containing over 20 guest rooms, roughly 300 out of the city's 450 residential hotels.

In April of 2004, a light fixture ballast fell onto plastic audiovisual storage containers and other combustibles in an unoccupied storage room of a California hotel. The resulting fire in the loaded storage room caused $3,500 in damage to construction and $5,000 in content damage. Prior to the arrival of the fire department, the fire was put out by a single fire sprinkler. Without it, there would have been sheer havoc in the occupied three-story hotel.

Hotel chains develop their own corporate policies, drawing heavily on the model building codes, and most make life safety a top priority. Their responsiveness to fire prevention and safety issues, particularly since 1980, has been comprehensive and steady. This attitude is admirable, and everyone benefits. Particularly in high-rise hotels, the powers that be have recognized that fire sprinklers are fundamental to occupant safety, and some chains have been proactively installing sprinklers in their existing hotels without being forced (by legislation) to do so. Their reasoning addresses the fact that although fire is a low-probability occurrence, it is an event of high consequence. The bottom line is that the U.S. lodging industry has seen a steady decline in fires (there were 12,200 U.S. hotel and motel fires in 1980) and in fire-related fatalities (approaching 75% over the last 30 years), which is due primarily to stricter codes, a decrease in cigarette smoking, and an increase in public awareness.

It is worth noting that improved fire protection system technology has catalyzed itself in the lodging industry. In a fire-safe hotel, one activated sprinkler sounds an alarm, and only a few are necessary to stop the fire before it can produce an abundance of carbon monoxide-carrying smoke. From 2005 to 2007, 82% of fires occurring in hotels and motels were confined to the room of origin (the leading causes of the larger fires were arson, heating equipment, and electrical malfunction). Costs for retrofitting older hotels are high, however, the use of CPVC pipe (in lieu of steel) has made things reasonably affordable. The advantages of using plastic pipe include lowered costs for material and labor, and a slick C-factor of 150 to provide hydraulic advantages, which means smaller piping. On the downside, CPVC has a lower temperature resistance, fittings are not reusable, and the piping is more susceptible to damage and vandalism. But the installations are a sound dependable wellspring of safety. Modern development of aesthetically tasteful (concealed) sprinklers makes the final product one that is markedly subtle.

Building codes require a Fire Department Control Station installed in large hotels. The responding fire officer in charge will use this alarm panel to monitor building controls, transmit voice instructions, and manually operate certain smoke control devices. By anyone's standards, interconnected smoke detectors should be placed in all rooms of the hotel. These will alert guests and staff to the threat of fire before sprinkler activation, and (ideally) will be tied directly to the local fire department. Smoke alarm effectiveness data compiled by the National Fire Incident Reporting System notes that in hotels and motels, smoke alarms alerted occupants in 86% of all fires.

It all boils down to overall hotel fire safety, which may include the following features: fire extinguishers, manual pull-boxes, standpipe systems, emergency lighting, exit signs, gas supply shut-off devices, pressurized stairways, duct smoke detectors, fire-resistant construction, smoke control systems, adequately sized exits, and staff emergency response plans. There is no valid reason for any doors to be propped open and inspectors are aware of this quality control issue, particularly for stairways. Doors should be able to automatically shut and latch tightly.

It is paramount that a written plan be established that delineates responsibilities for hotel employees in case of an emergency. What is of considerable concern for management is that building occupants not be exposed to false (nuisance) alarms, because the last thing they want to do is to inconvenience their guests while they are enjoying a climate of comfort. Regular inspection and maintenance of alarm systems is a must. Unless inspected and tested, fire safety equipment may be of no good use. Although NFPA standards dictate the frequency of these inspections, it is prudent to inspect and test older fire equipment more often than the newer devices, and document activities accordingly.[1]

Statistically, the areas of fire origin within a hotel are most commonly bedrooms, laundry rooms, and kitchens, in that order. Cooking is the cause of 46% of all hotel and motel fires, but 97% of those are very small, resulting in little property loss. Electrical malfunction is the number one cause of the larger fires which may involve the entire structure. The incidence rate of hotel and motel fires peaks from 6:00 to 9:00 p.m., with a secondary peak between 8:00 and 9:00 a.m. Overall, hotel structure fires are more apt to occur on weekends.

The following housekeeping tips should be fundamental procedure:

- Do not empty ashtrays into anything containing other materials.
- Unplug coffeepots in guest rooms when not in use.
- Have furnaces regularly cleaned and serviced.
- Eliminate the continued use of any unsafe electrical or heating equipment.
- Do not overload electrical outlets.
- Report unusual appliances in guest rooms such as space heaters or hot plates.
- Empty vacuum cleaner bags on a regular basis.
- Secure premises to protect against potential arson attempts.
- Ensure that all exits and exit routes are not blocked in any manner.

Fire studies have shown that fire sprinkler systems are the most reliable part of any building's fire safety plan. It fits the bill. Listed fire equipment must be tested at recommended intervals. Building personnel must be fully instructed in the intent and purpose of the various fire alarm signals, so that an informed and qualified individual can step up and respond skillfully should the time arise. Everyone has heard of past fire tragedies, and no one wants to see history repeat itself in that regard.

19

Multiresidences

Many cities, most recently Atlanta (in 1989), Philadelphia (in 1991), San Diego (in 1992), Sarasota (in 2000), and Chicago (in 2005) have passed various forms of legislation requiring some degree of fire sprinkler protection within high-rise residential buildings. Fires in buildings over 75 feet high are not manageable incidents because firefighting and rescue operations are unable to reach the upper stories of a building from the ground. Some high-rise retrofit ordinances do not apply to residential high-rise buildings, many of which are many present in urban areas. Retrofitting high-rises with fire sprinklers is an extremely expensive upgrade, costing over $5 per square foot in most cases. Those opposed to the retrofit ordinances argue that condominium owners and renters simply cannot afford this amount of money. In addition, the retrofit installation of fire sprinklers in older apartment buildings and co-ops would expose walls and floors, which may necessitate expensive asbestos-removal projects. To control costs, some cities allow the installation of CPVC plastic pipe in residential high-rises, or mandate fire sprinklers in corridors and public spaces (only), with the addition of one sprinkler poked inside the entry door of each tenant unit. What flies under the radar here is the silent threat to condominium owners and apartment dwellers, which is the danger of a neighbor's fire, perhaps spreading during the middle of the night. See Figure 19.1.

An apartment building fire will quickly fill the unit with smoke before the arrival of any firefighters. Smoke will spread quickly. The National Fire Protection Association (NFPA) reports that although 94% of flame damage in a high-rise apartment building fire is confined to the room of origin, only 58% of smoke damage from the same fire is confined to the room of origin. The room or area of origin in an apartment fire is most frequently the kitchen (41%), followed by the bedroom (16%), and the living room, family room, or den (9%). The following are fire prevention facts with regard to high-rise residential structures:

- Fires in unsprinklered buildings are very hard to control.
- A high-rise fire is a firefighter's worst nightmare.
- An abundance of fire is intensely hot, choking, and life-threatening.
- Interior firefighting is delayed by the distances that firefighters have to climb through black smoke, while carrying heavy equipment and breathing apparatus.

FIGURE 19.1
The aftermath of an apartment fire contained by the activation of one residential fire sprinkler (photo by John Koski).

- The large number of people that are inside a high-rise building renders it impossible to expect an immediate evacuation under emergency conditions.
- Ninety percent of all fires in sprinklered buildings are controlled by fewer than four activated sprinklers.
- There has never been a multiple loss of life in a fully sprinklered building.

In the event of a fire within a high-rise structure, the golden rule is to respond quickly without panicking. Never use the elevator unless such use is authorized by a fire fighter. Proceed to the closest stairwell and walk down without attempting to pass a cluster of others. Remember that the best air is always close to the ground, so stay low.

It is advisable to schedule fire drills for all high-rise multiple residences or, at the very least, maintain an updated emergency action plan for all residents. In high-rise fires, firefighters must often cope with logistical problems trying to locate a fire, while struggling with evacuation issues when windows are sealed and thus inaccessible from fire department ladders. Also, smoke travel will increasingly impede egress attempts in halls and stairwells. If a fire does occur, it will be more manageable at a time when no one is asleep.

Dormitories and Fraternity Houses

We are familiar with tragic cases of college dormitory fires. What is immediately noticeable in a dormitory walkthrough is the tremendous combustible load. Two students and their belongings are crammed into 250-sq ft living spaces. There is a lot of "fuel" there to burn. Another cause for concern is that your typical dorm-room occupant is not your garden-variety, safety-conscious responsible adult. There are a lot more smokers. There are candle- and incense-burners. There are more overloaded electrical sockets than you'll ever see in one building. And if you want to become acquainted with every kind of hotplate and small cooking appliance known to modern man, just visit a dormitory. Add to this the fact that for a third of the day you have a building filled with sound sleepers, and you have the formula for potential disaster. Fortunately, colleges across the nation have been (slowly) retrofitting these facilities with fire sprinkler systems for the past two decades. But this progress has not been fast enough to satisfy fire protection officials.

The NFPA reports that over the past 30 years the same fire protection deficiencies have been noted in all fraternity house fires. In that time, all of the very destructive fires have been in fraternities as opposed to sorority residences. There are several underlying reasons why this fact is substantiated by fire statistics. For one, there is more alcohol consumption in fraternity housing. Fire and alcohol are a bad mix. Sleeping occupants will very likely be unaware of a developing fire in the building, and are sometimes badly disoriented and confused when they are finally awakened. If they have already inhaled carbon monoxide and other toxic fire gases by that time, they may not be alert enough to be able to locate a route by which to crawl out into a safe atmosphere. The typical fraternity house fire originates in a public area and soon spreads into the sleeping areas. Firewalls rarely separate the two. See Figure 19.2.

Fraternity houses are often of older construction (many date from the 1920s), and most contain wide open staircases. Housekeeping practices and general building upkeep are not up to par with that of sorority houses. Numerous smoke-filled parties often lead to fraternity members removing batteries from smoke detectors. Fraternity buildings are often privately owned and are more likely to be situated off campus. Like other independent campus housing, these usually are lacking in fire-rated construction and self-closing corridor doors, and have sleeping areas that are not separated from public rooms. "Nonsanctioned" fraternity houses are a problem. These can become multiresidences when a member's father floats a long-term loan to a band of Greeks. But it is usually the characteristically lax behavior of occupants that increases the chance for fire loss in a fraternity house. This list is typified by rough treatment of the building, greasy kitchens, space heaters, cigarette smoking, emptied fire extinguishers, large-scale parties, trash

FIGURE 19.2
A garage fire may spread to other areas of a multiresidence if not extinguished in timely fashion by firefighting personnel (photo by Steve Bittinger). (See color insert following p. 52.)

and combustibles stored at exits, dead bolts on doors, and frequent overnight guests who may not be familiar with the exit routes within the building. Arson is more common to fraternities, and dorms are not exempt from this menace either. A 19-year-old Murray State University student lost his life in a 1998 arson fire in an unsprinklered dormitory that housed 300 students.

Small fraternity houses are just as likely to burn as the large ones. Although the general rule is that larger properties are more likely to be sprinklered than small ones, you still are faced with the fact that due to the type of construction often present in these structures, there exist combustible concealed spaces where fires can start or spread. Older houses have interior walls consisting of wood lathe covered by plaster. Worse, fraternity bedrooms and living spaces are notorious for interior wood paneling finishes, often installed on furring strips that are nailed to the plaster walls. This added "fuel" will proportionately intensify a fire.

The National Fire Sprinkler Association's (NFSA's) Sorority/Fraternity Retrofit Task Force has assembled a one-hour risk management and awareness program that has been presented to the Fraternal Executive Association, the National Fraternity Organization, alumni associations, and other involved groups. Among other valuable information they provide, the task force is emphasizing that fire sprinklers will protect "even those who are too intoxicated to exit a burning structure." A bill requiring 16 University of Wisconsin system residence halls to be retrofitted with automatic fire sprinkler systems came on the heels of similar legislation passed in the state of Virginia following a fire that killed one student. At the public hearing, (Wisconsin) State Representative Rob Kreibich commented that "parents are in essence handing their kids over to the state, expecting a safe and secure environment." State fire sprinkler retrofit ordinances passed in 1998 include Michigan (all dormitories) and Texas (state university high-rise dorms). Illinois governor Pat Quinn signed a bill into law in July 2010 that required all sorority and fraternity houses in Illinois to have an automatic fire sprinkler system. Existing fraternity and sorority houses have a 9-year window of compliances and the law states that noncompliance means no occupancy.

Engineering design criteria for the Wisconsin dormitories, which are all over 60-ft tall, will be based on NFPA pamphlet #13. NFPA #13R would apply to all dormitories, apartments, and other campus housing of four stories and less. The NFPA #13R system allows economic consideration by allowing the omission of sprinklers from various areas including attics, small closets, and bathrooms. NFPA #13D is rarely used for dormitories, but may be used when certain firewalls and firestopping measures make for two-unit compartmentalization. The #13R and #13D codes also allow for the installation of CPVC piping, a substantially easier retrofit installation. (See Figure 19.3.) Quick-response residential sprinklers must be used. These heads, developed in the late 1970s, are considerably more sensitive to heat and thereby quite effective in terms of controlling residential fires before small rooms can fill with toxic smoke. Some extended-coverage residential pendent sprinklers are capable of covering a 20 ft × 20 ft area per sprinkler. When NFPA #13 systems are used, economies are realized by using the quick-response sprinklers, which gain a hydraulic advantage with the (code-sanctioned) allowance of a reduced design area. NFPA #13R is the more cost-effective standard, with life safety as its primary goal, property protection being a secondary peripheral objective.[2]

In a fraternity, half-way house, or dormitory, the norm is to use sidewall sprinklers on piping that should somehow be inconspicuously concealed. After piping is installed, custom soffits may be installed for that purpose. Or if the areas to be protected are not "highly decorative," then the piping may be painted and left exposed. The cost-effective approach there is to have the pipe shop-painted prior to installation. Engineers must be cognizant of the specific fire hazards present in all areas of the building. They must be on the lookout for unheated areas and also for combustible blind spaces that must be protected with sprinklers. They should seek to minimize the number of necessary wall penetrations, particularly at masonry walls.

In multistory buildings, the reality of getting the pipe in place must be addressed. For high-rises, delivered pipe lengths should be kept to 12 ft in length, as they must be transported via stairs or a freight elevator, and then fed through ceiling grids to be wiggled between structural members for proper placement. When ceiling tiles are to be removed and replaced, the

installer can expect that many tiles will be soiled or broken, and should figure on buying any number of new ones before the project even starts.

Colleges and universities were not built in a day; the age and design of
each building varies considerably. As a consequence, multiple buildings may
contain different fire alarm system types, all reporting to the same central
monitoring station. If all these systems cannot be networked together, maintenance and upkeep compatibility will become a major issue. Many campus buildings are older, however, and the lifespan of a fire alarm system is
much less than the lifespan of the building housing that system. Code compliance should be achieved during renovation of older buildings to resolve
that issue, as well as issues regarding egress insufficiencies, change of use,
and any construction-type deficiencies. All dormitories ensure the safety
and security of students by protecting the entry of unwanted intruders with
devices on the outside of the structure. Firefighting personnel must possess
a required "key" for entry during emergencies.

The driving force behind requiring fire sprinklers is the current building
and fire codes. Because there is undisputable evidence that automatic fire
sprinkler systems can effectively protect both life and property, continued
strenuous efforts must be made toward mandatory code adoption of sprinkler retrofit ordinances nationwide for campus housing. A 1 am fire in a
New Hampshire dormitory was extinguished (in December of 1998) by fire
sprinklers. This blaze, in Holloway Hall at Keene State College, originated

FIGURE 19.3
Prior to the installation of a drywall ceiling, CPVC sprinkler piping is routed through ceiling
joists so that its presence may be completely concealed.

in a third-floor trash room where discarded smoking materials had ignited nearby materials. When firefighters arrived, they found that the sprinkler system had doused most of the flames. The dormitory had already been evacuated, and firefighting crews worked to prevent building damage of any significance. Due to the early hour of this fire, there is no doubt that the in-place sprinklers saved human suffering and prevented fire spread. Approved fire sprinkler design and installations for existing structures must be a primary objective of universities to ensure a safe living environment for all college students.

Endnotes

1. Mark Bromann, "In a Climate of Comfort," *PM Engineer,* August 2005, pp. 20, 22, 24.
2. Mark Bromann, "The Dormitory and Fraternity House Fire Problem," *PM Engineer,* April 2000, pp. 26, 28, 30.

20

Enclosed Shopping Malls

The enclosed shopping mall is largely an American innovation that arrived on the scene in the late 1960s. The covered area of the typical enclosed mall has a wide pedestrian "public" walkway connecting buildings that house single or multiple tenants. An arrangement of three to seven "anchor" department stores combined with a plethora of small specialty retail outlets comprise the tenant docket within the climate-controlled environment. Almost all malls are of one- or two-level noncombustible construction. They are open and "airy." Some covered malls are built with attached parking decks, restaurants, recreational and entertainment areas, movie theaters, food courts, video arcades, customer service offices, and other amenities. For the most part, tenants operate retail stores whose contents represent the major potential hazard. Storefront closures tend to be rolling overhead metal grilles that can be fully recessed, a shopper-friendly design that maximizes retail sales by encouraging customer flow. This open walkway-to-store design also serves the need for speedy occupant egress.

A shopping mall falls under the category of "multiple merchandise stores." An average mall containing 100 spaces plus anchor stores, covering about 700,000 sq ft, will have a value exceeding $60 million and will have contents totaling somewhere in the vicinity of $100 million. Some enclosed malls are extremely large facilities, and may exceed one million sq ft in area. The two largest, the Mall of America in Bloomington, Minnesota and the West Edmonton Mall in Alberta, Canada, can be classified as "megamalls." The West Edmonton Mall boasts over 800 stores, including 11 large department stores and over 100 eating establishments. Housed within its 5.2 million sq ft are other attractions including an indoor water park, an NHL-sized hockey rink, a zoo, a 360-room hotel, art exhibits, and a nightclub area. It opened in 1981.

The Mall of America, open to the public since 1992, contains an endless array of stores surrounding a seven-acre amusement park complete with roller coaster and Ferris wheel. It also contains a large aquarium, home to sharks and other marine creatures. Its elevators service four levels and a basement. Many hotels are nearby, providing shuttle services to the mall. These superstructures are more than malls; they are vacation and tourist attractions.

The Mall of America hosts more than 39 million visitors per year, creating a special fire protection objective of life safety. Thousands of people expect to come and go safely each day. Due to the wide-open expanse that malls feature, smoke control is just as important a design factor as the fire protection systems and the egress layout. The common mall promenade areas are open atria that span from floor to roof, making smoke accumulation a major concern. The Mall of America features a pressurization and smoke control/evacuation system that controls the spread of smoke through the regulation of air movement. Railings and glass shields surround the vertically aligned floor openings. Their system includes 185 exhaust fans and 21 air handlers capable of circulating 105,000 cu ft of air per minute. The smoke detectors and related dampers are tested annually, and the mall's four emergency generators are tested twice each month.

Plans detailing smoke removal systems must detail the type of exhaust/ventilation system as well as locations of air intakes. Water curtains provided around floor openings also act as smoke barriers. These are part and parcel of the sprinkler/standpipe systems. All three model building codes dictate that 2-1/2-in. fire department valves be installed at mall entrances and exits, at entrances to egress tunnels and passageways, and at enclosed stairways (that open directly into the main shopping area). You may also find hose valves with 100 ft of hose placed (sometimes) in odd locations, and this is designed intentionally so that fire department personnel can reach any area of the structure with a 30-ft hose stream. The (IBC) International Building Code requires that covered mall buildings, "[B]e equipped with Class I hose connections connected to a system sized to deliver water at 250 gallons per minute," at (Section 905.3.3) the following key locations:

1. Within the mall at the entrance to each exit passageway or corridor
2. At each floor-level landing within enclosed stairways opening directly on the mall
3. At exterior public entrances to the mall

Sprinkler and (if required) standpipe systems are zoned per area and (it is hoped) paralleled with fire alarm and smoke zones so that any signal to the addressable fire alarm panel can accurately pinpoint the source of the smoke, heat, or fire sprinkler activation. Well-organized monitoring is a critical facet that serves as an additional incentive to achieve the goal of a comprehensive safety plan acceptable to all authorities having jurisdiction over large commercial public buildings. Zones typically separate tenant areas, backroom service corridors, and mall common areas.

The high turnover of retail tenants in shopping malls means that the engineers must focus their attention toward considerations for future rework within the building. The usual approach is to provide mall management with hydraulic information (available water pressure and volume) for each tenant space water supply feed for fire protection. They should be individually identified, and their corresponding water flow supply data are to be made a permanent record for reference. Although it is unlikely that the supply source and related piping to the tenant flanges will be changed, the piping downstream of the valve and flanges may be altered time and again. Most existing malls make use of this efficient engineering arrangement, a strategy that speeds the process by which new tenants can open for business. Tenant space design criteria are normally .20 gpm/sf over 1,500 sq ft when supplied by wet-pipe systems.

Some malls have their own water sources whereas others take their water from city water mains. Some have pumping stations, some have individual fire pumps, and some have no need for any fire pump at all. It all depends on the water pressure and volume characteristics that are available. If a fire pump is present, the common areas of the mall make an excellent application for light-hazard extended-coverage pendent sprinklers. These may be used in any event where calculations prove out their hydraulic adequacy.

In the interests of life safety, some malls dictate that all fire sprinkler modification work be done between 10 pm and 7 am regardless of system design. Depending on the arrangement of valves, some systems require a drain-down time of up to two hours. These problems can be alleviated with the addition of sectional valves (previously mentioned) for each tenant space. Tamper switches, as well as ongoing testing and maintenance, will ensure that the problem of accidental valve closure is practically nonexistent.

Two-hour rated stairwells and exit passageways are designed for protection of shoppers and are funneled along paths of egress. Emergency lighting and fluorescent signage is often provided in these areas, as is smoke control, conditions favorable for fire-safe exiting. All three model codes require a one-hour fire resistance rating for walls between tenant spaces in order to limit the spread of fire. Building Officials and Code Administrators (BOCA) and Uniform Building Code (UBC) require this separation wall to be rated only up to the underside of the drop-ceiling, whereas the SBBCI Standard Building Code mandates that the one-hour rated separating wall be constructed from "slab to slab," or tight to the underside of the roof deck.

It is estimated that over 75% of the United States population visits a shopping mall at least once a month. There is potential for these high-traffic occupancies to experience dangerous life safety situations. Occupants come in all ages, with varying degrees of ability to move themselves quickly when egress action becomes necessary. Grade-level access is not always a step or two away. Thorough planning and coordination among architects, engineers, code authorities, and fire officials is the best foundation for overall shopping design strategy. And it must be noted that statistical documentation and case histories lay proof to the fact that fire sprinkler systems remain the most effective means to combat fire in all structures housing mercantile occupancies.[1]

Endnote

1. Mark Bromann, "Design Considerations for Enclosed Shopping Malls," *PM Engineer,* February 2007, pp. 12, 14, 16.

21

Hospitals and Healthcare Facilities

If uncontrolled, fire spreads quickly, producing an abundance of blinding toxic smoke. If you think long and hard about each and every building in your town, the very last place you would want a fire to start is in your local hospital.

Hospitals are large places. Virtually everyone who has visited a friend or relative in a hospital has had a similar experience after leaving the visiting room: difficulty finding the way out. And this is under normal conditions. Consider others in the same hospital trying to exit or just move to a safe location under emergency conditions: patients who are sound asleep, patients who are not able to walk normally, those sedated, those in treatment, those in intensive care, the very old, and newborn babies. A hospital or healthcare facility is most assuredly the type of standing structure in which we would never want to see the onset of fire.

Certainly, architectural and design features of fire safety have evolved over time and fire statistics have recorded a matching measure of success. Many hospital and healthcare facility fires are so small, and extinguished in the incipient stage, that local fire departments are often unaware of those occurrences. With an increase in fire safety awareness, buffeted by stricter code enforcement, the number of such small unreported fires has most likely risen over time. The number of hospital fires reported to U.S. fire departments saw a steady drop from 8,000 in 1981 to 3,500 in 1988. Adding all healthcare facilities (excluding nursing homes) to these totals, fire departments reported an unwavering drop from 13,000 to 6,600 fires throughout the same time period. Reported-fire figures have seemed to settle on a sort of low plateau. The 1999 figure for total structure fires in healthcare facilities was 3,400, resulting in 124 civilian injuries and no fatalities. There were a total of 190 civilian fatalities resulting from hospital and healthcare facility fires in the period from 1980 to 1999. However, despite the potential for fire within these buildings, when compared to other U.S. occupancies and structures the overall fire record for health care facilities is remarkably low. This is most likely due to the high numbers of staff on hand, fire-resistive construction (interior fire-retardant building materials), and the fact that combustible fuel load is very low in these facilities.

According to a 2006 National Fire Protection Association (NFPA) report, 5% of fires occurring in health care facilities extended beyond the room of origin. In addition to human error—resulting in doors propped open—walls may have unsealed openings that allow the passage of heat and smoke. Systems responsible for these openings (through firewalls and smoke barrier walls) include medical gas piping, ductwork, communication wiring, and HVAC ductwork. Inspectors should be unrelenting in their duty to check for firewall integrity.

Cause and Origin

Today, roughly one-fourth of all healthcare facility fires begin in kitchens or rooms otherwise set aside for food preparation. Thirteen percent start in patient rooms, and about eight percent of these fires originate in laundry areas. Fire, however, can ignite almost anywhere. Table 21.1 illustrates how the causes of reported fires have changed over a 20-year period in hospitals and healthcare facilities.

Less prevalent causes of origin are almost too numerous to list. They include generators, air-conditioning equipment, various types of heating components, televisions, refrigeration equipment, incinerators, and other items of medical equipment that are particular to hospitals, such as oxygen tanks. As improbable as it may seem, a number of fires in healthcare facilities have started when a patient simply didn't move far enough away from an oxygen cylinder before attempting to light a cigarette. Improper storage or transfer of liquid oxygen, unsafe usage of tanks in kitchens and workshops, and smoking around oxygen (by patients or visitors) are real risks for healthcare agencies. They are not limited to just the patient rooms.

A 1998 Missouri hospital fire started in an intensive care unit during surgery, when privacy drapes were ignited by an electric (2,200° F) cauterizing pen. This fire was quickly put out via the use of a chemical fire extinguisher. A fire that occurred in February of 2002 at the Fairbanks Memorial Hospital in Alaska was caused by a leak in a pressurized hydraulic oil hose in the incinerator room. One 286°F sprinkler in the room activated and contained the fire while alarms activated. Responding hospital staff soon completely doused what flames remained with the use of several fire extinguishers. An October 1996 kitchen fire at the Cape Cod Hospital (in Massachusetts) was

TABLE 21.1

Causes of Reported Fires

Cause Category, 1980		Cause Category, 2000	
Smoking	32%	Cooking equipment	28%
Arson (intentional)	14%	Arson (intentional)	13%
Cooking equipment	10%	Electrical systems	11%
Electrical systems	8%	Smoking	8%
Dryers	4%	Dryers	6%

effectively extinguished by automatic fire sprinklers in quick fashion. These unexpected outbursts of fire could have had far worse outcomes.

Code Compliance Debacle

Structures housing this nation's healthcare operations are among the most heavily regulated in our history. To keep code compliance simple and straightforward, what should be followed is a blend of NFPA #101 and NFPA #99. NFPA #101, the *Life Safety Code,* tells the reader when fire suppression and fire detection systems are required. It also addresses fire protection require- ments for occupant loading and egress requirements. NFPA #99, the *Health Care Facilities Code,* is a standard that determines the performance criteria for health-care facilities. It outlines degrees of risk for specialized areas within healthcare facilities such as audiometric booths, walk-in freezers, incinera- tors, compact shelf storage, rubbish chutes, elevator machine rooms, and also provides protection requirements for flammable liquids and gases.

However, when designing for a new hospital or new addition, the number of code documents that may require investigation are far too numerous to men- tion. At last count, there are over 60 NFPA documents alone that are applica- ble to healthcare facility occupancies. As the traditional healthcare landscape changes, the odyssey of code compliance becomes increasingly political. It's a code-saturated playing field. The typical U.S. hospital must comply with local fire and building codes, state fire and building codes, various require- ments scripted by state health department officials, one of the current model building codes, NFPA #101: *Life Safety Code,* the Americans with Disabilities Act (ADA) design guidelines, Healthcare Finance Administration (HHS/ HCFA) life safety codes, NFPA #99, and the design guidelines of the Joint Commission for the Accreditation of Healthcare Organizations (JCAHO) Plant and Technology requirements. This is a short list, which doesn't include any interim policy changes implemented by the above-captioned agencies. Pending location and local jurisdiction intervention, the time spent conduct- ing preconstruction, midconstruction, and plan review meetings will vary considerably. In any case, documentation of such meetings and strategy ses- sions will mount into a veritable paper jungle.[1]

Code requirements that are unique to the hospital setting will affect where doors are situated, the type of spaces that can open to a corridor, the exact size, height, and mesh style of privacy curtains, signage, building subdi- vision, size of storage alcoves, interior finish material limitations, level of required detection, fire-resistive wall ratings, automatic sprinkler and stand- pipe requirements, aisle widths, and just about anything else you can think of. The engineer, after determining exactly which agencies have jurisdic- tion, must make contact with those AHJs (Authority Having Jurisdictions) to verify what specific codes apply. It is not with great frequency today that hospitals are built brand new from the ground up. Instead, hospitals are substantially refurbished or earmarked for modestly size additions. If renovations are major, with significant changes in occupancy, these projects may well affect the existing structure, resulting in further instances of submittals for multiple code reviews. Walls may be relocated, water supplies changed, freezers added, changes made in storage locations and arrangements, and there may be a substantial increase in the number of persons employed in certain parts of the facility. When new equipment is purchased, it will probably be housed in an existing space. This equipment must be hazard-assessed, because some may produce excessive heat or flammable vapors.

Egress

Multiple facility additions will result in corridors that may be extremely difficult for patients and visitors to navigate. Hospitals built in the 1950s, 1960s, and early 1970s were of the high-rise variety, resembling tall hotels of that vintage. In all cases, residents are grouped closely together, and their visitors come and go at all hours of the day. One key difference though, is that hospital patients are categorized as those incapable of ordinary self-preservation. Special life safety design features are necessary inclusions for high-rise "towers." Protection of vertical openings is essential, due to the propensity of fire to travel upward with great velocity.

Procedures for egress in hospitals actually differ in sharp contrast from most other occupancies. The workforce employed in larger facilities is well prepared for emergencies, ready to efficiently relocate occupants and staff to safe areas within the building. This "defend-in-place" philosophy works well in light of the presence of active fire suppression systems and other measures previously undertaken within the building for fire and smoke containment.

Fire Suppression and Equipment

A recent hospital inspection revealed no less than 13 intermediary control valves spread willy-nilly along a feed main, which were apparently left by the installers of automatic sprinkler system additions of yesteryear. These valves are "time-bombs," serving no real purpose and only responsible for increasing the chances that a valve may someday be accidentally shut. The inherent lack of record keeping, and the failure to retain and store old fire sprinkler plans, worsens this problem. If record keeping is poor, sprinkler systems provided for future hospital additions will typically add more large feed piping than is really necessary, adding undue structural load in older sections of the building. The accurate preparation of hydraulic calculations may also be adversely affected. The real Achilles' heel in this mix is the failure to maintain an organized system of safety records. This holds true for all building systems. The design of engineered air-handling systems that automatically contain smoke in zoned areas is an equally complicated task that must come under careful scrutiny when buildings are modified.

The major suppression equipment necessary for health-care fire safety can be summarized and prioritized as follows: (1) automatic fire sprinkler protection, (2) automatic detection, (3) smoke control, and (4) fire extinguishers. Building compartmentalization for smoke and fire is a prerequisite. There are more than a few cases in which fire protection equipment has been found not to be fully operational, so staff must be capable of red-flagging such conditions. Fire pumps, sprinkler systems, smoke control systems, fire alarm systems, tanks, generators, and special extinguishing systems must have extensive, established inspection procedures. Maintenance and periodic suppression component testing is paramount, of equal importance in the hospital climate to education, programs, and fire drills.

Hospital settings change as technology improves to meet the changing needs of health care. Renovations, although tricky to coordinate, are nonetheless made necessary by these advancements. Fire statistics prove that fire safety and engineering design efforts to meet these challenges have been quite successful. But fires still happen in healthcare facilities at a rate exceeding nine per day in the United States. Nearly all of these fires are well contained. The lurking enemy, as always, is complacency. Particularly in a quiet, serious, and sterile environment, John Q. Public is pretty much unfazed by the potential for danger. But practically speaking, there is no such thing as a true "no-risk" situation. The fire service is only as good as its weakest link. Should equipment fail, or maintenance and inspection procedures falter, even the best building design is compromised. The practice and application of fire protection efforts within hospitals today is clearly a team effort.[2]

Endnotes

1. Mark Bromann, "Managing the Potential for Danger," *PM Engineer*, January 2007, pp. 16–17.
2. Mark Bromann, "Managing the Potential for Danger," *PM Engineer*, February 2007, pp. 12, 14, 16.

22

Nursing Homes

In these modern civil times, there is little doubt that the most distressing fire tragedies have occurred in nursing homes. Any student of history has been taught that societies have long been appraised by how well they care for and protect the most vulnerable members of society. There are presently over 15,500 nursing homes in the United States. They are home to many individuals who are not capable of self-preservation. Two nursing home fires in 2003 (in Connecticut and Tennessee) claimed 31 lives. In the aftermath, palpable questions were raised concerning nursing home life safety. The bottom line is that over 16% of all existing U.S. nursing homes today have no automatic fire sprinkler protection whatsoever. But it's more than that. It's also known that 25% of all nursing home fires occur in those not equipped with sprinkler systems (where the biggest danger looms), and that is an inauspicious and disturbing fact. This statistic cannot change without the adoption of strong retrofit ordinances. In the meantime, the level of fire safety in elderly care facilities is substantially compromised.

A grim modern-day disaster occurred on May 20, 1980 in Kingston, Jamaica, when 146 (of 211) residents died from fire in a home for older adults. The most devastating nursing home fire in the United States remains the February 1957 blaze in the three-story Katie Jane Memorial Home in Warrenton, Missouri that killed 72 of their (total) 149 residents. Faulty electrical wiring was the probable cause of that rapidly spreading fire. A headline in the *St. Louis Dispatch* read "Screams of Elderly Patients Were Quickly Stilled by Flames." The Missouri legislature met the day after the Warrenton fire, introducing a bill that required sprinkler systems for all Missouri nursing homes and similar institutions. Once again, existing codes were recognized as being antiquated only on the heels of a major catastrophe.

A case in point is an eight-story nursing home in Evansville, Indiana, which is devoid of fire sprinklers in corridors and patient rooms. While considering the addition of sprinklers, the owner received cost estimates exceeding $1 million for a retrofit installation. Barring any new code legislation, it is doubtful that money will be used there for fire sprinklers in the near future. One of their administrators commented that "We've never had a fire they couldn't contain with just a regular fire extinguisher." The Evansville Fire Marshal responded by saying that "It's an older building, multiple stories. It's not sprinklered, you've got elderly residents who are difficult to evacuate— all the combinations for a high-level disaster."

TABLE 22.1

Fire Sprinkler Systems in Nursing Homes

State	Number of Licensed Nursing Homes	Total Fully Protected with Fire Sprinkler Systems	Percentage 100% Sprinklered (%)
Nebraska	286	200	70
Pennsylvania	962	635	66
North Dakota	83	53	64
Minnesota	433	238	55
Colorado	220	112	51
Arkansas	225	110	49
New York	1,044	407	39
Utah	100	38	38
South Dakota	146	53	36
Michigan	409	147	36

The statistic most often cited is that of the 15,500+ American nursing homes, about 3,500 lack automatic fire sprinkler systems. According to the National Fire Protection Association (NFPA), over 2,300 nursing home fires are reported each year. Fortunately, most of these are contained quickly without injury or loss of life, and success stories abound.

The (typical) irony inherent to this discussion is that the older nonsprinklered facilities (those least resistant to the threat of fire) are exempt from most state and federal legislation mandating the inclusion of sprinklers. So the ominous potential for future tragedy remains, at least in certain states. At present, all nursing homes are equipped with full fire sprinkler protection in four U.S. states: Arizona, Vermont, Alaska, and West Virginia. There are a total of 325 nursing homes in those states, where about 6% of their collective population is over the age of 75. The other side of the coin is illustrated in Table 22.1.

Again, when fire occurs in a nonsprinklered facility that houses many who are disoriented or otherwise unable to evacuate competently on their own, there is a real risk of peril. The headlines read "Lack of Sprinklers Cited in Deadly Nursing Home Fire" after a 2006 tragic fire in Nashville, Tennessee that killed 8 women and critically injured 16. Nationwide, a total of 24 nursing homes have experienced fatal fires to some extent since 1999. That statistic, in this day and age, is inexcusable.

In a welcome trend practiced by numerous municipalities, fire departments are being trained in responding to calls from their local nursing homes so that they will be able to take appropriate actions when the time arises. Similarly, the training of all nursing home staff is critical to fire safety planning. Those on site at the time of an emergency must be ready to perform a series of tasks including (1) closing doors to rooms close to the fire area, (2) relocating residents, (3) meeting the arriving fire department personnel, and

(4) supervising residents and providing them with information regarding the extent or status of the emergency.

Over one million disabled and elderly Americans live in assisted-living facilities. Like nursing homes, they provide round-the-clock supervision, but they fall into a separate classification because they do not carry a 24/7 medical staff. There are 36,000 such licensed facilities in the United States, and there are thousands more that are currently unlicensed. Here are the dossier facts: over 700 fires are reported in assisted-living facilities each year. Even so, four states (Montana, Minnesota, Hawaii, and Massachusetts) set no fire safety standards at all for that industry. This jeopardized state of affairs, and other glaring gaps in safety regulations, need to change as there have been over 60 fatal assisted-living fires across the nation in the past five years. That's one per month, in structures inhabited by the general public.

Is Your Loved One Reasonably Safe?

When checking on a relative or friend already confined to a nursing home or assisted-living facility, reassurance of safety is not a guarantee, but can be gauged by observing the following:

- Look for safety amenities such as smoke or heat detectors, marked exits, fire extinguishers, and fire sprinklers.
- Are there actively followed guidelines for those residents who smoke?
- Does the facility appear to be well maintained?
- Are there any large items blocking exits?
- Do the patient beds have casters?
- Are hallways and corridors at least 8 ft in width?
- Is the staff instructed in terms of emergency egress planning?
- Is the kitchen clean and uncluttered?
- Are exits ample, with ramps and doors that allow for direct access exiting?
- When asked, does management respond definitively to questions regarding life safety?
- What is the staff-to-patient ratio? Is this ratio consistent throughout the year?

There are 1.4 million Americans residing in nursing homes today, and that number is expected to exceed 1.8 million by the year 2020. Fire is inevitable; life is transitory. But as everyone knows, effective measures can be implemented to protect the public from fire. Why are U.S. fire statistics measurably and consistently higher than those in countries overseas? One reason is that so much here revolves around money. That, coupled with grandfathering provisions, allows older substandard nursing homes to slide around code reform. They simply cannot afford necessary life safety upgrades and continue to stay in business. Although an assiduous gain, measured in lives, can be realized by retrofitting our older nursing homes with built-in fire sprinkler systems, the cost to do so has been estimated at close to $1 billion. That translates in dollars to about $715 per patient. It is hoped that there is still something called goodwill. If the trust and confidence of the paying customer is to hold any merit to the providers of a skilled nursing setting, they are the ones whose mission it should be to champion fire safety and petition their congressmen to introduce bills that will create funding to share the burden of costs for fire safety upgrades. Their input would carry a necessary influence, and the time is long overdue.[1]

Endnote

1. Mark Bromann, "Nursing Homes Have Sorely Needed Changes," *PM Engineer,* April 2006, pp. 24, 26, 28.

23

Storage Warehouses

Much of National Fire Protection Association (NFPA) #13, the *Standard for the Installation of Sprinkler Systems*, elicits installation requirements for the protection of stored commodities. Particularly in Chapters 12 through 20, the standard tells the reader the exact criteria by which to design a sprinkler system, and those criteria are based on the type of storage, storage height, and type of commodity being stored. The minimum installation requirements are based on years of extensive large-scale fire testing, and as a result everyone concerned can realize a sense of confidence that the installed sprinkler system will reasonably protect warehouse property against a potential fire. Because as sprinklered storage testing cannot realistically be replicated using computerized fire models, real data has had to be derived from tests that are representative of actual conditions. Thus, the magnitude of fire growth for various storage situations can be researched and quantified academically in conjunction with the fire suppression capabilities of differing levels of fire sprinkler protection.

Although most warehouses are of masonry construction with metal roofs, we know that warehouses will continue to experience fires. Regarding fire sprinkler systems, the keys to good storage protection involve more water from the initial operating sprinklers along with a quick sprinkler response time. Storage facilities in general can consist of such a wide diversity of hazards and a wide variety of products stored in a single building that they present one of the most challenging occupancies for those engineers whose responsibility is fire prevention. Within large warehouse structures, tenants may change after several years. The new tenant will likely be storing product in a different fashion, and may be housing products of a higher degree of fire hazard. To correctly apply building and fire codes, factors to be considered include the expected storage height, the sprinkler temperature rating, aisle separation distances, product encapsulation, the use of in-rack sprinklers, stable or unstable piles, and the type and methodology of storage. Newer warehouses built today have become increasingly large and some can be correctly classified as distribution centers, ranging in footprint size anywhere from 300,000 to 600,000 sq ft.

Arrangement of Storage

A common method of product storage is *palletized storage*, in which commodities stored on pallets create horizontal spaces between tiers. Air contained in these spaces would be quite welcome if you were building a bonfire. The horizontal spaces between stored combustible products, however, bode badly for fire protection because the potential for a hot spreading fire is much greater there than if the application were "solid pile." NFPA #13 defines *solid-piled storage* as that which consists of "storage of commodities stacked on each other."

Rack storage also creates horizontal spaces between tiers of storage. Racks are structured as single, double, or multiple-row. They may be portable (not fixed in place), which means that they are capable of being configured differently over time. They may contain vertical or horizontal barriers. They may be "open racks" or they may contain solid shelving. Depending on the protection criteria dictated by the NFPA #13 standard, they may include in-rack fire sprinklers, which will provide the majority of fire sprinkler performance in the event a fire starts in such a storage area. Widely scattered pallets of materials or other disorganized storage arrangements make system design difficult, but do not necessarily increase the hazard or potential risk. (See Figure 23.1.)

The flue spaces within rack storage configurations are not to be overlooked. Initially, fire will advance up these spaces very rapidly, which is important for timely sprinkler activation. The flue spaces will also provide for a generous degree of water distribution, controlling the fire while prewetting stored products that the fire has yet to reach. The critical maintenance item to maintain is clear flue spaces, both transverse and longitudinal, to ensure that they are not blocked in any way.

Sprinkler System Protection Options

Traditionally, fire sprinklers protecting *high-piled storage* (defined as storage of product exceeding 12 ft in height) are arranged in a system "grid." With this piping layout there are no real "dead-ends" of piping. Instead, all smaller "line" piping (directly supplying water to the sprinklers) is connected at either end to a larger "main" pipe. This way, any activated fire sprinkler receives water flow from two directions, providing a hydraulic advantage. The (larger) main pipe runs may also be "looped" together, further increasing water flow and optimizing water volume and pressure. When this isn't enough, or when conditions are of an overly hazardous nature and the existing water supply is insufficient when compared to the system demand requirement, a fire pump will be installed to augment the water supply and boost the pressure delivered to the fire sprinkler system.

FIGURE 23.1
Fire roars through an empty wooden crate stacked within a storage arrangement. (See color insert following p. 52.)

The fire sprinklers used in the sprinkler systems will vary depending on required flow. We know from test results that the downward velocity of a water droplet from a standard 1/2-in. orifice ($K = 5.5$) sprinkler will probably be overcome by the very strong fire plume produced by a high-challenge fire. The higher starting pressure of the first few activating sprinklers will result in smaller droplet sizes which may evaporate quickly. A high pressure discharge of water dropping toward a powerful fire may also tend to cold-solder an adjacent sprinkler, causing a "skip" in sprinkler coverage which also weakens fire penetration. To remedy this situation, larger orifice spray sprinklers are usually employed to knock down fire intensity and limit exposure.

The K-factor of a sprinkler is simply the relationship between pressure (in pounds per square inch) and volume (in gallons per minute discharged). The K-factor equals volume (gpm) divided by the square root of the pressure. So if we were to measure a discharge of 43 gallons per minute from one sprinkler, and we knew that the system was supplying a pressure of 60 psi at the elevation of that sprinkler, we could then ascertain that the sprinkler in question has a K-factor of 5.55.

The placard affixed to the sprinkler system riser will display the design density for that system. In a high-piled storage warehouse, that system design may be 0.40 gpm/sf over a certain design area, an area that has been forecast to suppress or control a fire should all sprinklers within that area activate. From the placard (usually a metal sign) information, we know that system will be able to deliver .40 gallons per minute per sq ft from each sprinkler in the design area. Design density always affects the choice of sprinkler to be used. For our example, many sprinklers may be acceptable options, but the sprinkler best suited and most economically viable would be one having a 14.0 K-factor. Table 23.1 shows the sprinkler types most suitable for a variety of required design densities.

An early suppression, fast response (ESFR) sprinkler may possess a higher K-factor such as 22.4 or even as high as 25.2. These sprinklers are often used in very tall distribution centers, designed to deliver a large amount of water onto burning commodities at the early stage of fire development without the aid of in-rack sprinklers, even up to storage heights of 45 feet.

TABLE 23.1

Favorable Sprinkler Types for Specified Design
Densities

K-Factor of Sprinkler	Design Density (gpm/sq ft)
5.5	Less than 0.20
8.1	0.20–0.33
11.2	0.33–0.38
14.0	0.38–0.44
16.8	0.44 and higher

What distinguishes the ESFR sprinkler from all other sprinkler types is that it is specifically designed to extinguish a fire (using very large water densities) rather than simply control the blaze until firefighter arrival. A sprinkler system utilizing ESFR sprinklers allows warehouse owners some flexibility with regard to future tenants, and solves operational difficulties by nullifying the installation of in-rack sprinkler systems. Accompanying this concept, however, is an abundance of code requirements which limits the circumstances under which they can be installed. Writers of the code are cognizant of the fact that any obstructions to ESFR sprinkler discharge (ductwork, large light fixtures, structural steel) may inhibit proper suppression performance and may thereby "lose the battle" against a high-challenge fire. ESFR sprinkler installation requirements are more "fussy" in nature and thoroughly covered in Section 8.12 of NFPA #13 (as one example, they may not be used in buildings having a roof slope of 2:12 or greater). But their effectiveness in test fires is unparalleled, due primarily to the density and momentum of the water they will discharge before a fire can develop.

> An in-rack sprinkler has been defined as an ordinary-temperature-rated upright sprinkler having either a 1/2" or 17/32" orifice size, designed for placement within storage racks, and having a metal "water shield" appliance affixed above the deflector that collects heat, protects the sprinkler head from damage, and prevents water discharging at a higher level from wetting its operating response elements.

Although NFPA #13 design criteria fluctuate depending on the hazard, storage style, and storage height, in-rack sprinklers are an effective means for stopping or controlling fire in its early stage of growth. Discharged water from in-rack sprinklers will quickly travel to burning commodities, absorbing heat in the process. Face sprinklers may be also used as an integral part of the system design. These are standard sprinklers located in transverse flue spaces or along the aisle, implemented to protect the external face of storage. Aisle width measurements are also considered when determining system design parameters. Greater aisle separation distances will lessen the potential for heat transfer and subsequent fire spread.

All area/density design curves noted in NFPA #13 are minimums, and in certain cases interpolation is permissible. It's ironic that loss prevention is one of the few professions in which overall success is measured by failure rates. For this reason, a safety factor (or *cushion*) is built in to all design density curves depicted in the NFPA standards.

Depending on the degree of hazard, another important component of the overall fire protection scheme for rack storage warehouses is a smoke removal system. Whatever specific type of smoke evacuation system is employed, it must be capable of removing smoke from the building automatically, and also by manual activation. Standard exhaust fans spaced around a warehouse are not to be considered an acceptable substitute for a high-powered mechanical air-handling exhaust system, if required. When properly installed, ventilation operations will protect occupants from heat and toxic fumes while giving rescue personnel the visibility necessary to locate the fire.

Hazards and Risks

What comes first and foremost in sprinkler system design is the classification of hazard. Most stored commodities are classified by category in Chapter 5 of NFPA #13. Aside from the common product classifications, certain commodities will require special design criteria that are outlined in special sections of the code. These include banned roll paper, rubber tires, plastics, aerosols, flammable liquids, archived material, tissue, carpeting, and idle pallets. The fire sprinkler system must be able to vigorously meet the fire challenge of whatever commodity is stored. Burning plastics represent a particularly high challenge because they burn hotter and faster than wood-based materials. What may fly under the radar is the fact that there are numerous stored products (housewares, auto parts, appliances, padded furniture, etc.) that are really a mixture of plastics and other materials. Stored plastics that are "exposed" are much more hazardous because when there is more "surface area" available for flame exposure, the advancement of fire growth will be intensified. By comparison, cartoned plastics will burn at a slower rate. Free-flowing plastics (those that will fall out of their containers during a fire) such as pellets, powder, or small plastic objects, are considered to be less hazardous storagewise, because they tend to "spill" and will thereby smother a developing fire. The evaluation of plastics and their storage arrangement must be done with precision, as an abundance of plastics poses a highly precarious threat. Sound judgment applied for these cases is extremely necessary to mitigate or erase the fear of a total loss from fire.

Specific products noted in NFPA #13 range from low hazard (Class I commodities) to the more hazardous (Class IV commodities). Examples of Class I commodities include bagged cement, metal file cabinets, canned food, electric motors, and bagged salt. Class IV commodities include fabric, asphalt shingles, oil-based paints, fiberglass insulation, glue, rubber, photographic film, and furniture with plastic coverings. All classifications are contingent upon the type and volume of material and its primary packaging. These classifications, however, are also affected by the amount of air space that is made necessary by warehousing and storage particulars. When classifications are not a clear call, the NFPA #13 code (Section 1.5) stresses that "nothing in this standard is intended to prevent the use of systems, methods, or devices of equivalent or superior quality, strength, fire resistance, effectiveness, durability, and safety over those prescribed by this standard." Products that are not targeted specifically for code conformance in the NFPA standards are addressed in other insurance company pamphlets and publications such as the *FM Global Property Loss Prevention Data Sheets*.

Idle (empty) pallet storage presents a severe fire condition of which most people are not aware. Whether they be constructed of wood or plastic, sprinkler system design is heavily "beefed up" to protect the areas where they may be stored. It is wise in any case to limit storage of idle pallets to a height of six feet. When stored outdoors, basic outdoor storage practices must first give realistic consideration to the capabilities of the local fire department. Regardless of how far away their fire station might be, the local authority will typically require a detailed outdoor storage plan that includes locations of fences and gates, fire hydrants, extinguishers, and the preplanned aisles and roadways. Paved outdoor storage yards will often have painted lines outlining storage areas that have been laid out on the submitted plan. The areas used for outdoor storage should be clear of weeds, plants, and vegetation that will become dry at some point and thus become fuel for fire spread, or possibly even a source of fire ignition. To avoid any fire exposure, adequate clearances to adjacent properties are a must, particularly in coastal areas prone to high winds. And as with indoor storage, general "congestion" is a friend of fire that paves the way for its growth and spread. All commodities should be stored as low in height as possible. Long narrow piles are preferable to large square piles of material. Other advisories for outdoor storage include:

- Keep the site free of unnecessary combustibles.
- Lock up the site overnight.
- Have fire extinguishers readily available.
- Smoking should be prohibited.
- Maintain minimum aisle widths of 15 ft to allow for fire vehicle transport.
- Keep pallet piles separated by a minimum of 30 ft.
- Prohibit welding and cutting operations in the storage area.

What fire sprinklers cannot ever hope to comprehensively control is a fire caused by an explosion, a rupture fire that creates a fireball, or any fire involving severe "jetting." These fires produce large volumes of heat release that can overwhelm a sprinkler system. During a fire, for example, a 55-gallon drum containing a flammable liquid may rupture, which will ignite its entire contents all at once. If many of these drums are stacked on pallets, an explosive atmosphere certainly exists. For storage situations of this severity, fire sprinkler systems (or AFFF foam protection) are nonetheless required and there are several NFPA pamphlets that cover special design criteria. NFPA #30, *Flammable and Combustible Liquids Code*, outlines many specific provisions and guidelines for recommended storage practices for these high-hazard situations.

In an organized workplace, virtually any facility can be transformed into a very safe place for business. For all storage conditions, regular inspections are a key to safety. The bottom line is that the facility manager must prepare some type of emergency plan, and keep that plan current. Communications regarding fire safety must be made available to all employees. Cooperation with the local fire department is also very essential. All incidents of an emergency nature are unique, however, the response of facility personnel will play a major role regardless of the incident at hand. Personnel working in a storage warehouse must be well aware that fire protection and safety are top company priorities.

24

Correctional Facilities

Oily rags ignited a blaze on the roof of the overcrowded three-story Ohio State Penitentiary on April 21, 1930, resulting in a horrific fire that left 322 inmates dead from smoke inhalation. The undetected fire erupted minutes after iron gates had closed 4,500 men to confinement in their cells. Not a single guard or staff member had been trained in how to respond to a fire. The jail's warden was initially certain that the fire had been designed as part of an escape plot. The fire advanced prodigiously, becoming so hot that a tower of catwalks warped and twisted into a snarl of metal. The 40-year-old structure contained six tiers of cellblocks, and those on tiers five and six were trapped by smoke and flame. Most inmates were eventually evacuated despite the ensuing chaotic nightmare.

That prison fire in Columbus, Ohio remains the deadliest in U.S. history. On its heels, the National Fire Protection Association Life Safety Code (NFPA #101) and other model codes called for new jails to be constructed of limited- or noncombustible materials, and to be provided with automatic fire sprinkler and detection systems.[1]

But until 1977, few automatic fire sprinkler systems were present in correctional facilities. In June and July of that year, three major fires in unsprinklered prisons ended in a total of 68 fatalities. In St. John, New Brunswick, 21 perished in their cells due to smoke inhalation following an arson fire. Just five days later, on June 26 in Maury County, Tennessee, 42 died. As with the New Brunswick tragedy, this one was incendiary, and a lengthy search by staff for lost keys was noted as a major contributing factor. Five more fatalities occurring on July 7 in a federal prison in Danbury, Connecticut, initiated public awareness and nationwide sympathy for those dying inhumanely while locked in a cell. Gradually, smoke detection and fire sprinkler systems were accepted as sensible and practical inclusions to correctional facilities.

Current data reveal that property losses decrease by 86% in prison fires where sprinklers are present. The chances of dying in a prison fire are cut by two-thirds if the prison is sprinklered. Still, the NFPA reported that in 1999, only 28% of correctional facility fires occurred in areas that were protected by automatic sprinklers.

Living behind Locked Doors

Among all occupancy groups, correctional facilities may be the most difficult to protect from fire, in part because the cardinal rule of immediate evacuation does not apply. Life safety code requisites are invalidated in contradictory fashion by heightened security measures which require that all fire exits be locked, chained, or completely obstructed. To resolve this conflict, jails adhere to a "protect-in-place" strategy, opting to relocate inmates from the area of fire origin to a secure area within the facility. This is safe standard operating procedure. Currently, 91% of all U.S. inmates are male. Their average sentence is 53 months, and their average age is 36. Most correctional facilities will be in operation for about 40 years without any major renovation.

Not surprisingly, fires occur all too often within U.S. jails and detention centers. Seven out of ten are intentionally set. In the 138 British prisons, 674 of the 980 fire occurrences during 2003 were recorded as ones of "malicious ignition." Fortunately, most arson fires in jails are quickly extinguished. In case you're wondering, it's not realistic to expect anytime soon that inmates who smoke will be denied access to matches or lighters. Authorities estimate that half of all prison fires go unreported. Any one of nine basic triggers factor into any incidence of inmate arson:

- To intimidate other inmates or staff
- As a distraction to assist others attempting to escape
- As a means of revenge
- To draw attention
- As a means of suicide
- To protest living conditions
- To relieve the boredom of incarceration
- As a show of force during a prison riot
- When inmates with severe mental problems "act out"

Riots are a real threat. During a seven-hour New Brunswick conflict on June 17, 2007, rioting inmates broke lights, ignited one fire, smashed windows, and shattered fire sprinklers in common areas. When facilities are overcrowded, the task of successfully moving or evacuating individuals becomes more arduous. And an increased prisoner density within confined spaces breeds a greater propensity for malice. In addition to deliberately set fires, the leading causes for prison fires include cooking and heating equipment, smoking, electrical system malfunction, and clothes dryers. Of all laundries in the United States, one in six reports a fire each year, with dryers to blame in 70% of those instances.

Comprehensive Safeguards

As more new prisons were equipped with automatic sprinkler systems throughout the 1980s, and more existing jails were retrofitted with sprinklers, authorities took note of an increase in the inmate suicide rate. Those suffering from mental illness or emotional distress posed an obvious risk. Such a person may start his bedding on fire to attempt suicide, an action that would certainly place the lives of others in peril. In California, home to the nation's largest state prison system, sprinklers themselves had been used as an anchor for numerous hangings. Although some facilities retrofitted in sprinklers, they refrained from placing them in housing cells to preclude suicide attempts and the possibility of self-injury. Institutional sprinklers were developed as a result, designed to be quite sturdy yet possessing the capability to "release" from its frame with a suspended pull in excess of 50 pounds.

The term "tamper-resistant" takes on a whole new meaning inside jails, where some inmates are prone to do almost anything to disrupt normal operations to interrupt their monotony. Sprinklers in the 1980s were viewed as an irresistible temptation for vandals. Another fear was that inmates might disassemble the sprinkler heads and use the parts as weapons. Institutional sprinkler technology responded by developing heads with durable tamper-resistant components that are capable of a short-angled flush installation. Brass-body sprinklers such as the Tyco Model TFP MAX Institutional Horizontal Sidewall or the Reliable Model XL Institutional Pendent have a proven track record, and are mainstays in mental institutions and prison applications today. Form precedes function. Especially for areas where inmates may be unsupervised, fire sprinklers and their cojoined escutcheon plates must be able to withstand attempts at tampering.

Design Is the Key

In 1990 I was involved with document preparation and the preliminary retrofit sprinkler system design for an occupied 29-story maximum security prison. Focusing on the work itself was usually not difficult, one distraction being the disclaimer to be signed each day which stated in so many words that people held hostage were pretty much on their own. The total security concept is meticulously strict, causing long delays when moving from one area to another. Eventually I was able to access all areas except the armory, a room housing emergency ammunition.

What can burn? Beyond the armory, kitchen, library, carpentry shop, mechanical/electrical rooms, and the laundry, cell blocks are a haven for personal effects: clothing, papers, computers, magazines, mattresses, and radios.

One inmate had part of his cell occupied by a stack of pornography that was four feet high. Small television sets, books, and blankets round out the mix of combustibles. Many jails use fire-retardant cotton bedding to reduce the fire potential.

Design meetings were attended by eight to ten staff members who scrutinized each nuance of design. The first item nixed for this concrete structure concerned the use of Hilti guns, strictly *verboten* for the installation of hangers. Sealants were to be carefully specified for all soffits and escutcheons, so that it would be impossible for anyone to hide a razor blade, drugs, or other contraband between the concrete and flush metal coverings. The goals established included immediate containment and control of fire and ensuing smoke.

For all prisons, a primary concern is prompt fire detection to alert both staff and occupants of peril, along with zoning to efficiently pinpoint the source of a fire. The favorable design approach utilizes small independent sprinkler zones, allowing staff to quickly identify the location of water flow. This is critical from a fire protection standpoint while also managing and mitigating water flow resulting from vandalism. When personnel are able to shut down isolated sections of the sprinkler system following sprinkler activation, all other areas of the sprinkler system can remain in service to protect the jail.

For the aforementioned federal prison, smoke detectors were guarded with 22 gauge mesh security housings that had adequate openings for smoke entry. It was decided that if these detection devices could not be reasonably protected, the secondary backup option was to install the smoke detectors inside HVAC system ductwork only.[2]

Other noteworthy amenities included:

- A Class I combination standpipe servicing one stairwell, providing 65 psi at the uppermost outlet
- Exit-stairway pressurization systems to activate upon water flow
- Main control room surveillance and notification
- Central station fire department notification
- Systems divided by zones for each podular housing area, each with its own control valve, tamper, and flow switches
- A smoke-control system timed to depressurize the fire floor
- Electrical supervision of all fire sprinkler system devices
- A fire command station complete with all fire pump, control panel, and annunciation indicators
- Standby power support for all emergency and lighting systems

Budget and space considerations prohibited the very sagacious alternative of double-interlocked preaction systems to protect the structure, but the wet-pipe scheme serviced the need for quick failsafe water application. Preaction systems are utilized when it is ascertained that sensitive electronic equipment contained within sprinkler-protected areas must have little chance of any future exposure to water. As with any project, the designer must ask the questions: what potential problems are unique to this occupancy? What may be encountered here that is appreciably diverse from the norm?[3]

Something not to be overlooked is the search for a contractor experienced in the litany of details, applicable codes, installation delays, and construction methods to be implemented for correctional facility system projects. Overall success hinges on an experienced professional team effort, and a willingness to agree on long-term life safety goals.

Endnotes

1. NFPA #101 is authored in part by a technical subcommittee on Detention and Correctional Occupancies and contains one full chapter outlining specific required provisions for prisons.
2. The goal is to make equipment, piping, and ductwork inaccessible to inmates while remaining accessible for future maintenance work.
3. Mark Bromann, "Fire Protection for Correctional Facilities," *PM Engineer,* February 2008, pp. 18, 20.

25

Vacant Structures

Numbers don't tell the whole story, but U.S. fire statistics are meticulously tabulated and sadly, quite factual. Every year, approximately 31,000 fires occur in U.S. structures that have been previously coded by fire departments as vacant buildings. The primary suspect in each vacant property fire is automatically an (unknown subject) arsonist, since over 70% of these fires have historically been classified incendiary or suspicious. Sometimes the firesetter's actions have not been deliberate. Homeless people have been known to start small fires in vacant buildings as a source of heat, that one way or another spread and burn out of control. Statistically, "child-play" starts fires in abandoned buildings at a rate three times that of the occurrence for child-play fires in school properties.

There are more vacant commercial buildings out there than you would think, currently over 250,000 in the United States.[1] This figure is currently on the rise. The properties in question tend to be small. Sixty percent of vacant buildings are between 1,000 and 5,000 sq ft, and 89% of all vacated structures claim less than a 10,000-sq-ft footprint. These buildings are older: 87% of them were erected prior to 1969 and 90% of them are one- or two-story structures. They may be empty factories or they may be deserted apartment houses. They all offer easy access for juveniles, serial arsonists, vandals, and the homeless.

Technically, an *idle* structure is one in which normal business operations have been temporarily halted. The business manager must articulate to staff a catalog of special fire protection safeguards for such a property, to ensure that the building is clean and its fire protection systems (sprinklers and alarms) are maintained in working order. A minimum temperature of 40°F should be maintained[2] and the local fire department must receive notification that the building is temporarily unoccupied. A *vacant* structure has been defined as "a property in which normal operations will not be resumed and from which most of all the production equipment and furnishings have been removed." There are no serviceable fire sprinkler or alarm systems inside. The structure is truly vacant.

On August 15, 2002, an abandoned dwelling caught fire around midnight in Jersey City, New Jersey. The heavy fire soon spread to adjacent occupied homes. Not only were properties destroyed, but seven families were left homeless. The property loss numbers for vacant building fires are staggering. A 1998 oceanfront fire in a wood-frame Alaska seafood processing plant burned for 45 minutes before the fire department was called. An electrical short was determined to be the cause of the blaze in this 26,000 sq ft, two-story edifice, which had been vacant for about 18 months. Although there were no fatalities, the dollar loss amounted to $5 million. Vacant building fires result in millions of dollars in property damage in the United States each year.

An annual average of 11 civilian deaths and 66 injuries are the direct result of vacant U.S. building fires. Perhaps the most tragic risk posed by abandoned buildings is to the firefighters themselves. In an abandoned structure fire, the chance for firefighter injury triples. The larger the building, the greater is the risk. A vacant run-down warehouse in Detroit, known to be frequented by vagrants and derelicts, experienced several small fires extinguished by firefighters in 1987. A fireman was killed in a more severe, subsequent inferno in the same property when it unexpectedly flashed over. The fire spread to an adjacent property, where two more firemen perished as a result of a wall collapse. See Figure 25.1.

In December of 1999, a vacant warehouse building fire in Worcester, Massachusetts induced firefighters to take the offensive because a homeless couple was known to use it periodically for shelter. When some of the firemen lost their bearings inside the burning building, others rushed to their aid. Tragically, six of the firemen lost their lives in this effort.

There are 1,800 all-professional fire departments in the United States, protecting 44% of the population, primarily in the nation's cities. Firefighters are

FIGURE 25.1
The potential for sudden wall collapse or major wall crumbling is a life-threatening reality faced by all personnel fighting large structure fires (photo by Mike Charnota).

FIGURE 25.2
Windy gusting conditions greatly increase the chance that a vacant structure fire will ignite an adjacent property (photo by Mike Charnota).

very cognizant of safety procedure, but theirs is an extremely hazardous occupation. Because of the heightened and sometimes hidden dangers, firefighting procedures for vacant buildings are completely different than those for occupied structures. See Figure 25.2. One of the objectives of the fire department is to extinguish fires in deserted properties, but this obligation is not primary or mandatory. Life safety of the firefighters is always a key consideration. The method of employed attack is a decision that must be made expeditiously by the commanding officer, and many factors come into consideration for his imminent agenda including the size and height of the burning building. The CO will also need quick answers to the following questions:

1. Is the building properly secured?
2. How does the local water supply match up with the expected fire flow needed to mitigate the blaze?
3. Is there a possibility for exposure fires?
4. Will existing wind speed add to the exposure risk?
5. What is the extent of the present fire?
6. How reliable is the building's structural integrity?
7. For how long has the building been vacant?
8. Where is there sufficient street access?

Even if a fire sprinkler system is visible within the building in question, it is very often the case that the water supply for a vacated property has been shut off. The occurrence of a vacant building fire does not necessarily mean that there is no civilian life hazard, because children, the homeless, workmen, drug gangs, or random trespassers may be inside. Abandoned multiple dwellings represent the most severe exposure in terms of human life, and are not to be trifled with. Responding firefighters look for the following indications to ascertain the likelihood of squatters in peril:

- Curtains, plants, or window shades
- Extension wires strung from utility poles or adjacent buildings
- Lights on in any windows
- Signs of forced entry in a secured building
- Old vehicles parked outside

One very critical consideration is the building's structural stability, and if that is in question, the emphasis of initial fire attack is almost always from the exterior.[3] Any aggressive interior attack strategy must be physically feasible and within all known or perceived safety limitations. A vacant structure that has experienced previous fires might well be comprised of weakened (or burned-through) ceiling and floor beams, and related damages that give priority to precautionary measures over and above the normal goal of minimizing property loss. The initial advantage goes to the beast in a vacant building fire, because no working alarm system means a delayed response time by the firefighters. Efficient portable radio communication is essential.

A fire occurring in a vacant property is five times more likely to be an arson fire as compared to structure fires in occupied buildings. Low-income areas are the hardest hit, as are older inner-city areas. It has been estimated that 40% of these arson fires can be categorized as "arson for profit."

Obviously, arsonists have no concern for firefighter safety. It is not uncommon for a vacant building to be "rigged" in various ways to actually endanger the lives of those engaged in fire rescue efforts. This is about as sick as it gets, but fire departments realize that with a vacant building, an arsonist has had time to plan his insidious bonfire. Reported arsonist strategies in New York City have included loosened or removed bolts on fire escapes, refrigerators positioned so that they will fall on anyone entering the fire area, stairs removed, furniture blocking doors, and holes in floors covered over with cardboard or linoleum, to collapse under someone's weight. In addition, diesel fuel may be spread on all floors of the building. That way, a small initial fire may spread rapidly, trapping rescuers. Similarly, delayed ignition devices or accelerants may be strung from overhead beams, designed to drop into the fire as firemen advance. The list goes on, but suffice it to say that vacant building fires will always have a rapid fire spread potential due to the many interior shafts and exterior openings that create a substantial source of oxygen to feed the flames, with plenty of deserted combustibles and rubbish to burn.

Disadvantages notwithstanding, firefighters are conscious of the danger signs upon encountering vacant building fires, and the operation strongly endorsed to reduce destruction is by means of exterior attack. Fires that start in abandoned properties are a recurring menace, thus many fire departments maintain a listing of all vacant property within their jurisdiction to be monitored under a continuing level of scrutiny. Some cities implement signage "systems," which not only label vacant buildings but also provide vital information such as whether the building is structurally damaged, slated for demolition, or otherwise condemned. Fire department strategy strives to be one that provides rapid control while maintaining the least overall hazard to all personnel. It is work that is appreciated and applauded by the entire fire protection community.[4]

Endnotes

1 The number of vacant residential housing units in the United States has grown from 15.7 million units in 2005 to over 19 million in 2009. A spike in foreclosures (caused by the mortgage crisis) and sinking real estate values nationwide have been cited as the cause. In a weakening economy, there is a rise in vacant residential property numbers and more homes will be "burned for profit."

2 This temperature recommendation is per *FM Global Loss Prevention Data Sheet* 2-0, Section 2.4.2.

3 The use of tower ladders and aerial platforms is much preferred over traditional methods such as stretching out multiple hoselines inside a vacant structure for firefighting purposes.

4. Mark Bromann, "A Not-So-Innocent Danger," *PM Engineer*, May 2004, pp. 20, 21, 24.

26

Paint Spray Booths

In a new or existing building, certain words trigger an ominous condition in the eyes and ears of a fire protection engineer. Words such as: rubber tires, plastics, petroleum, inks, nitrocellulose film, aerosols, process vessel, oxygen tank, roll paper, propane, acetylene, foam rubber, dust accumulation, or polyethylene. This short list consists solely of hazardous commodities. Other words that jump out at fire protection engineers concern places where certain commodities may be found, that is, throughout rack storage, behind theatrical stages, in engine rooms, or within paint spray booths. The term "spray booth" is just what the name implies: it refers to a self-contained boxlike "booth" that physically surrounds paint spray operations. The most popular version is the "open front" arrangement that has a ventilation system used to control airflow to ensure that flammable vapors are removed to a safe location. Spray booths are large and awkward-looking metal contraptions, power-ventilated structures that have an innocent appearance about them that hides a very real fire threat.

What makes an atmosphere flammable is the oxygen in the air alongside a flammable gas, vapor, or dust in a volatile mixture. As with any area where flammable finishes such as airborne lacquers and paints are regularly applied, automatic fire sprinkler protection is a necessary component for life safety and property protection. In most jurisdictions, this requirement is enforced whether the building has an existing fire sprinkler system or not. Spray-painting operations pose a severe fire hazard, specifically in the specter of a rapidly spreading flash-type fire and a not-so subtle potential for explosion. Leftover residues from paint, finishes, and other flammable solvents serve to compound the risk, as these deposits are readily ignitable.

Automatic sprinkler protection must protect the booth interior, the area(s) behind the filters, and the interior of the exhaust duct. It is imperative that the exhaust duct be properly protected because it is a haven for accrued overspray deposits, which may lead to an occurrence of spontaneous ignition, and this ductwork is difficult to keep clean. The fire protection system design is fairly simple and originates with a feed main taken from the overhead sprinkler system. This line piping drops to an area on the side of the spray booth or an adjacent wall, complete with an accessible and supervised control valve and a 1-in. auxiliary drain. Appropriate signage (red lettering on a white background) should clearly identify both the control and drain valves.

FIGURE 26.1
This setup for a spray booth system controlling valve includes a 1-in. ball valve serving as an auxiliary drain. Signage for the control valve is missing as is the connecting alarm wiring for that valve's tamper switch.

It is not a bad idea to include a durable metal sign on the booth itself reading "Danger—Flammable Liquids." National Fire Protection Association (NFPA) #33, *Spray Application Using Flammable or Combustible Materials,* requires that the sprinkler system protecting the spray booth be designed for Extra Hazard Group II. Sprinkler spacing inside the booth must not exceed 90 sq ft, and inside ducts the sprinklers should be installed every 12 ft or less. A separate alarm is not necessary. See Figure 26.1.

Paint spray booths are a common presence within most automobile dealerships and body shops. If that building is sprinklered, its overhead system will be designed for an Ordinary Hazard Class II occupancy for the service areas, parts department, body shop, car wash, and similar areas. Rarely will a fire spread from the spray booth to other areas because the fuel contained within the booth is consumed very quickly, and the ventilation system continues to pull heat and flammable vapors away through the ductwork to the outside of the building. The whole idea is to sprinkler the spray booth, and to separate spray operations from other building processes in order to prevent ignition sources in either one from spreading and igniting fire in the other. Both must be designated "no smoking" areas. An adequate supply of portable fire extinguishers must accompany spray areas and mixing rooms.

In colder climates, the sprinkler protecting the exhaust plenum and ductwork is typically a dry pendent or dry sidewall sprinkler, a routine practice explicitly endorsed by the writers of NFPA #33. Accumulated loading or coating of fire sprinklers within the booth is to be avoided, so some methodology must be adopted in order to prevent paint spray from reaching the sprinklers. A thin plastic or paper bag tied around the sprinkler base is the usual cost-effective application. Polyethylene or cellophane bags, with a maximum thickness of 0.003 in., work best. If the sprinklers do become coated they must be cleaned or replaced. If the sprinkler is reasonably accessible, motor oil or Vaseline can be used to break down and wash off deposits. The temperature rating for these sprinklers inside the booth should be as low as possible; a rating of 135°F is recommended. The NFPA requires monthly maintenance inspections of all spray coating areas. An inspector must immediately red-flag any control valve closure and any instance of a sprinkler being coated with paint. All air-handling equipment filters should be cleaned on a regular basis.

No sprinkler system? No problem! In situations where water supplies are undersized or not readily available, dry chemical systems are frequently used to provide spray booth protection. See Figure 26.2. This configuration employs an actuation system to set off the nitrogen gas supply and trip the discharge of a dry powder agent, a combination of solids and gas, through dry chemical nozzles. We all know that baking soda will put out a fire. Similarly, the dry chemical agent smothers a blaze by interrupting the chain reaction of any flammable liquid fire. The dry chemical supply should be placed as close to the hazardous area as possible but outside the spray booth itself. Piping must be evenly distributed so that the system is balanced. Dry chemical nozzles come in different sizes and are of varying models, the "local" type being the desired choice for spray booths. Base code requirements and allowances are found in NFPA #17, *Dry Chemical Extinguishing Systems.*

CO_2 systems are another option, but one that is not often undertaken for spray booths. CO_2 attacks the fire by lowering the oxygen content. When O_2 falls below 15%, the fire goes out. However, this total flooding process takes a while to extinguish the fire. Combine that with the inherent life safety issue, and it's easy to see why a dry chemical agent or standard fire sprinklers are the preferred modes for spray booth suppression. The use of halon in spray booths, much more expensive but a method that does extinguish fire quickly, has been outlawed for quite some time due to the ozone-depleting nature of halogenized agents.

The installation of paint spray booths must adhere to the following (also see Figure 26.3):

- Spray booths should be constructed of #18 gauge steel.
- Flooring should be noncombustible and easy to clean.
- The exhaust duct should be sufficiently supported and run as directly as possible to the outside atmosphere.
- Spark-producing or open flame equipment used for welding or cutting shall be located in separate plant locations.
- No electrical equipment or switches should be located within the booth.

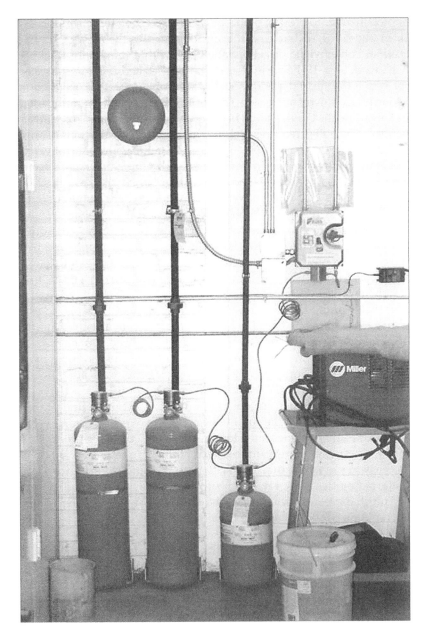

FIGURE 26.2
This dry chemical system adjacent to a large spray booth will operate effectively without a source of water for fire protection.

FIGURE 26.3
The electrical lighting for this paint spray booth is situated behind a rated glass panel that is sealed to prevent leakage of vapors or mist at its edges.

- Paint spray operations shall not be located below grade.
- All metal parts of the booth and duct must be grounded to prevent static charges.

There are no code parameters or cutoff points for the amount of flammable or combustible liquids used in spray booths, or for the frequency of their use.[1] But containment of flammable paints is certainly a key prevention strategy. Limiting quantities of flammable and combustible materials around the spray booth area will guarantee that a potential fire will be easier to manage, and will minimize ignition potential. If the system employed uses automatic fire sprinklers, the system must be hydraulically calculated per NFPA #13 code criteria, but that calculation need not include surrounding overhead sprinklers; they represent separate fire areas that are independent of the spray booth system. When both are properly protected by in-place fire protection, the risks and hazards posed by the presence of any spray booth can be effectively controlled, and sprinklers will have the final word.

Endnote

1. Mark Bromann, "Protection and Loss Control for Spray Booth Operations," *PM Engineer*, June 2007, pp. 18, 20.

27

Signage

Emergency signage has long been a staple of fire safety on the industrial landscape. The general rule for facilities is that approved standard identification must mark all exits with the exception of the main entrance. In addition, listed directional exit signs (at least 6-in. high per National Fire Protection Association (NFPA) #101) must appear in every location where the direction of travel to the nearest floor exit is concealed from view from any point of occupant travel. The traditional red and white exit sign may include an arrow. Because exits vary in identity, an almost unlimited array of exit signage has evolved into those including language such as ROOF ACCESS, PUSH TO EXIT, THIS DOOR TO REMAIN OPEN WHEN THE BUILDING IS OCCUPIED, IN FIRE EMERGENCY DO NOT USE ELEVATOR—USE EXIT STAIRS, FIRE DOOR—DO NOT OBSTRUCT, IN EMERGENCY SLIDE TO OPEN, NO EXIT, AREA OF REFUGE, and so on. Exit signs are typically illuminated and must be visible from any direction of exit access.

Identification signs for fire sprinkler systems are normally painted metal, red epoxy with white lettering, hung with a 3/16-in. nickel-plated steel chain. NFPA #13 mandates the use of signage primarily for the identification of valves and the determination of system locations that they serve. Section 6.7.4.1 notes that "all control, drain, and test connection valves shall be provided with permanently marked weatherproof metal or rigid plastic identification signs." Sections 6.7.4.3 and 6.7.4.3.1 read that "the control valve sign shall identify the portion of the building served. Systems that have more than one control valve that must be closed to work on a system or space shall have a sign referring to existence and location of other valves." *FM Global Property Loss Prevention Data Sheet* 2-0, under Section 2.6.5, advises to "Provide each Inspector's Test Connection with an identification tag that indicates the system being tested."

The General Purpose signs seen in valve rooms are 9 in. × 7 in. in size, governed by Section 7.7.1.5: "Caution signs shall be attached to all valves controlling sprinklers." The wording for these is expected to go as follows: "This valve controls fire protection equipment. Do not close until after fire has been extinguished. Use auxiliary valves when necessary to shut off supply to auxiliary equipment. CAUTION: Automatic alarm will be sounded if this valve is closed." The fact that the sign is actually present overrides any concerns about the exact wording asked for by the code. What is important is that the sign is clearly legible and its installation is such that it will remain

FIGURE 27.1

permanently affixed at the valve. The older General Purpose signs read, "This valve controls supply to AUTOMATIC SPRINKLERS, must be OPEN at all times, to be handled by authorized person or by employee caring for sprinkler system. In case of fire do not shut valve until fire is ENTIRELY OUT. When valve is shut for emergency repairs or fire notify :_____. Request directions. Restore protection quickly." Code requirements for all signage exist to avoid homemade signs containing embossed plastic tape or ink that may be "permanent" in name only. See Figure 27.5.

Drain and test connection signs are somewhat smaller (6 in. × 2 in.), red/white or white/red, and affixed to the valve. Some of the more popular 6 in. × 2 in. signs available contain the following wording (also see Figure 27.3):

Main Drain	Alarm Line
Auxiliary Drain	From City Main
Inspector's Test	Pump Discharge
From Pressure Tank	Standpipe Control Valve
Antifreeze System	Standpipe Drain
Tank Filling Line	Air Line

Many jurisdictions require specific wording on the door of the fire pump room or a room containing system control valves (see Figure 27.1). Such wording varies all over the board but conveys the message clearly, as does

FIGURE 27.2

FIGURE 27.3

the wording of the sign sometimes seen outside above the fire department connection. The Chicago Building Code reads that "a sign shall be placed near the outside bell where readily visible and shall bear the following: SPRINKLER ALARM, WHEN BELL RINGS CALL FIRE OR POLICE DEPARTMENT." Section 8.17.2.4.7.1 of NFPA #13 requires that "Each fire department connection to sprinkler systems shall be designated by a sign having raised or engraved letters at least 1 in. in height on plate or fitting reading service design- for example, AUTOSPKR, OPEN SPKR, AND STANDPIPE." See Figure 27.2, Figure 4.1, and Figure 6.5.

FIGURE 27.4

FIGURE 27.5

The code requirements exist so that system installations are held to an acceptable standard. Regarding the fire sprinkler piping itself, NFPA #13 (Section 6.3.8) requires that each piece in excess of 2-ft long be marked at the factory, so that the pipe type, model designation or schedule, and manufacturer can be identified. See Figure 27.8.

FIGURE 27.6

The most overlooked, most frequently misplaced, and in many cases the most important sign of all to be included in the original installation is the "hydraulic placard" or nameplate. Requirements governing these signs can be found in Section 24.5 of NFPA #13, and at least one placard should accompany each system riser. These come in all sizes (the typical ones are 4-1/4 in. × 6 in.) and should simply never be lost, because they contain not only the system design density particulars, but also the hydraulic demand (in volume and pressure) created by the installed fire sprinkler system. This information is vital for any future system additions or modifications, occupancy changes within the structure, retrofitting of backflow prevention devices, and any future change with regard to the water supply.

Acknowledging the reality that original sprinkler system shop drawings and hydraulic calculations are often lost or discarded, the NFPA has mandated (starting with the 2007 edition of pamphlet #13) that the installing contractor provide a "General Information Sign" (Section 24.6) which is to contain a plethora of design information including occupancy details, flow test results, low point drain locations, commodity and storage particulars, and in short, just about anything one would need for future reference regarding the installed system. I have yet to see such a sign and when I do, it will probably be in the form of some laminated long narrative, as Section 24.6 is very lengthy in its requests for information. Basically, this "sign" is intended as an information replacement backup in the event that the original shop drawings and calculations are either long gone or forever buried in a file somewhere.

FIGURE 27.7
An adhesive sign clearly marks this Schedule 10 black steel fire sprinkler system piping, while also indicating direction of flow.

The list of fire safety signs that may be contained in a building is far too long to discuss concisely. Suffice it to say that appropriate signage (of any number) is a necessity for the following conditions and situations:

Stairwells	Smoking Areas
Compressed Gas	Fueling Stations
Fumigants	Flammable Liquids
Fire Extinguishers	Fire Hose Stations
Occupant Load	Emergency Exits
Fire Routes	Cooking Equipment
Special Hazards Systems	Container Storage Cabinets
Explosive Materials	Portable Tanks
Elevators	Emergency Power Systems
Electrical Assembly	Carbon Dioxide Gas

Standard identification is about protection, safety, and substance. The lack of required signage should be a highlighted punch-list item on every fire inspection, particularly that for final occupancy. Signs exist for safe egress, for room labeling, to avoid confusion, for component or hazard identification, and to provide information crucial to inspectors, designers, engineers, and all building personnel.

28

FAQs

Q.: What is the rule with regard to fire sprinklers in elevator shafts?

A.: The Safety Code for Elevators and Escalators, ASME A.17.1, requires that upright or pendent sprinklers be installed at the top of all elevator hoistways. Under 102.2c, that code mandates that an automatic means be provided to disconnect the main line power to the elevators before water is delivered into the elevator shaft. That power shutdown can be accomplished by installing a detection system that will operate prior to sprinkler activation. The concept is to interrupt power before any discharge of sprinkler water. That could cause elevator malfunction, as could excessive heat. It could also wet the brakes of the elevator, rendering them unable to stop and hold a moving elevator. Section 8.15.5.3 in National Fire Protection Association (NFPA) #13 specifies that the temperature rating of the sprinklers in the shaft be no higher than 212 degrees.

In a fire scenario, smoke detector activation would recall the elevator to the ground floor (or to the lowest floor that is not experiencing activated detectors); then the main line switch would be automatically disconnected and the sprinkler would subsequently discharge. As noted in NFPA 101, "An elevator shall not be considered a component in a required means of egress."

A sprinkler at the bottom of the elevator shaft is only required if the shaft is combustible or contains combustible hydraulic fluids. The sprinkler at the top of the shaft may be omitted if the hoistway is noncombustible or of "limited-combustible" construction, and there are no polyurethane-coated steel belts or similar combustible belt material present. When a sprinkler is required, it must be independently supervised by a water flow indicator, which will transmit a signal when water does flow to that sprinkler.

The elevator shaft can definitely contribute to the unwanted movement of smoke and fire. The walls of the hoistway are almost always fire partitions having a two-hour fire-resistant rating. When hot smoke and gases accumulate at the top of the hoistway, a mushroom fire can result in the upper floor of the high-rise. For this reason, all of the model building codes require venting of all elevator shafts.

All firefighters are cautioned of potential significant dangers when using elevators during a fire. From an eight-year study of 178 high-rise fires in New York City, it was documented that elevators failed during one-third of the fires, and 37% of those failures were the direct result of water or fire damage to the elevator's electrical system.

However, the use of the building's elevators is of considerable help when battling a high-rise fire. Because of this, the design of elevator cars has become extremely sophisticated in recent years. The firefighter's strategy involves traveling in the elevator to a level that is three floors beneath the lowest floor that is reporting trouble. He ascends the stairs from there. It is imperative that the elevator be operable (without the sprinkler discharging) to attain peak effectiveness of firefighter efforts.

Q.: We are calling out white polyester-coated recessed pendent sprinklers for a project in Chicago. We were told that Chicago doesn't permit white sprinklers. Where can I find this in the code?

A.: There is nothing in the Chicago Building Code that disallows the use of white polyester-coated pendent sprinklers. The misconception stems from wording in the code that is related only to (fully recessed) concealed-type sprinklers. The wording for this is found in Section 15-16-130 and reads as follows. "Concealed sprinklers shall be equipped with cover plates which operate at a temperature of not greater that 135 degrees Fahrenheit. Such cover plates shall have a non-painted metallic, brass or chrome finish." In effect, what Chicago is prohibiting is the presence of any nonmetallic concealer cover. White, blue, brown, or any color is not kosher for the cover plate. What the writers of this code wish to avoid is someone painting a ceiling in the future, and painting the cover plate as part of the work. They are surmising that a (metallic) brass or chrome coverplate will be less likely to be painted. If the cover plate gets painted to the ceiling it will not be able to disengage in timely fashion if a fire occurs. This code amendment is wise policy, but is not incorporated as part of the NFPA codes.

Q.: Our local fire official is insisting that the fire sprinkler system also be installed above our swimming pool. Although I cannot follow his logic, I was wondering if the sprinklers above the pool won't corrode over time. Is this a valid concern?

A.: I have been involved in several instances where a local fire prevention office has waived this NFPA #13 requirement where about 90% of the floor area of a room is taken up by a swimming pool or ice rink. In those cases, what the local jurisdiction has accepted is simply a 3 or 4-in. pipe and cap stubbed into the pool area, in the event that a change of occupancy ever takes place. The thinking is that there is very little chance of a fire occurring in the pool area.
   That being said, you have some options with regard to the sprinklers. Because the moist atmosphere will accelerate corrosion, you would be better off with wax-coated sprinklers above the pool. Another option would be to install white polyester-coated upright heads in that area. Rusting may also

show up on the pipe itself, therefore galvanized pipe may be in order. What many contractors opt for in these cases is the installation of black steel piping that is shop-primed with tnemec paint. The zinc contained in this U.L. listed rust-inhibiting paint protects the pipe from moisture via a urethane film coating. The Series 394 tnemec coating can be sprayed on or roller applied, but it must be applied correctly. Any deviation from the recommended application process can lower performance and shorten the expected life of the pipe.

Q.: The bulk of our stored commodities are copper products, and we are naturally very sensitive about water damage. What precautions should we take before awarding a fire sprinkler system contract?

A.: Water damage from fire sprinkler systems is much more likely to occur after the work is completed and the sprinkler contractor has already packed up and left the job-site with the system in service. Your main responsibility will be to ensure that adequate heat is maintained in all areas of your facility. If a wet-pipe system is installed in areas of the building that may periodically have insufficient heat, then you are going to have trouble for sure. Each winter, contractors respond to a plethora of "freeze-up" service calls, typically after a thaw. That's when frozen water in a system expands as it melts, popping sprinklers and cracking fittings. Similar results may be experienced if a reduced-pressure (RPZ) backflow prevention device is an included component in an antifreeze loop, and the expansion chamber is of insufficient size or otherwise incapable of handling fluid expansion downstream of the RPZ.

Water damage has also occurred as a result of the following maintenance oversights:

- A sprinkler head subject to physical damage (installed in a gymnasium, warehouse, in racks, or at very low elevations) is whacked by a broom handle, forklift, or fused by a nearby welding torch.
- A dry-pipe sprinkler system was not properly drained following a system test or inspection service operation.
- The fittings on plastic pipe were poorly glued during installation. (Glued plastic fittings do not always have a long octogenarian life as one might expect. Incidentally, never test plastic pipe using air pressure, as it will probably shatter.)
- Dry-pipe sprinkler system piping was not properly pitched for drainage.
- Bulky incidental items were hung from fire sprinklers (hastily, banging the sprinkler).
- Press fittings on steel pipe were incorrectly turned or only partially seated.

- Insulation or freeze tape was not adequate to prevent cold drafts from freezing water in the system piping.
- Low-point auxiliary drains or drum-drips on dry-pipe systems were ignored for long periods of time

NFPA #25 requires that low points on dry systems be drained after each operation and before the onset of freezing weather conditions. Freeze-ups are your number one cause for concern. Insurance companies report that the average cost per water damage claim is $21,000. But the bottom line is that you're far better off with automatic fire sprinkler protection: the amount of water which is put on a fire by fire department hoses in an unsprinklered building fire is tens to hundreds times more than sprinklers would have discharged to accomplish the same task.

Q.: I service a system that should follow the NFPA #20 code, which says that the control panel should have two circuit breakers for any short circuit. Please tell me the reason for butting two circuit breakers, with nothing in between, especially when they are directly in series. If one trips then the other will be useless and not able to function. I know there is a technical reason for this, otherwise why can't we install just one circuit breaker?

A.: There really is only one circuit breaker. The other one acts as an isolation switch. It's the "second" circuit breaker that acts as the common breaker for the unit. When it becomes necessary to service that breaker, the (first one) isolation switch can be turned off for that purpose. If it weren't present, then the procedure would involve calling in Commonwealth Edison [the electric company] to disconnect power, which could mean an extended and very undesirable delay.

Q.: I'm trying to determine the temperature rating of fire sprinklers in our warehouse. They are all large-orifice upright sprinklers with a green bulb. There is no temperature noted or stamped on the sprinklers, so how do we figure this out?

A.: Yours are intermediate temperature-rated heads. The fact that the fluid inside the frangible bulb is green means that these are probably rated 200 or 212°F. A rating of 175° is also considered to be an intermediate tempera-ture rating, however, in that case the bulb color would be yellow (reference Table 6.2.5.1 in NFPA #13). In a fire, the heat-sensitive glycerin solution inside the bulb will expand, causing the bulb to shatter and thereby allowing water to flow from the sprinkler. The bulb's liquid is rated to temperatures as low as 65° below zero. Sprinklers with red (or orange) glass bulbs have an ordi-nary temperature classification and will activate at a ceiling temperature of (typically) 155°. Blue solutions in the bulb refer to high-temperature classification heads, which activate at around 286°. Purple solutions in the bulb are indicative of a sprinkler rarely used ("extra-high"), which will activate when ceiling temperatures reach 340°F.

Q.: We were provided a specification requiring black steel piping for sprinkler and standpipes. Joints are to be welded to ASME 31.1. This is raising a concern for the contractor. What specification does the majority of the critical buildings use for welding of fire sprinkler systems?

A.: The commonly accepted specification for pipe-o-let welds on sprinkler systems is ASME 31.9. You may notice that welds from most fabrication shops also feature a stamp. This identifies an individual welder, and the stamp on the pipe has been hammered on with a numbered or lettered punch. The cut-out (wafer) portion of the pipe will not accompany this piping, although in certain cases you may see this pipe "catalog" hung near a flow switch or some other device that required a hole cut into installed piping. It serves as visual proof demonstrating that a hole had been cut in the piping prior to the actual installation of the unit or weld-o-let, a validation not unlike the presence of a dead agave worm at the bottom of a bottle of tequila, authenticating some level of alcohol content.

Q.: Management states that cardboard boxes stored in Inergen-protected areas are to be removed. They claim that Inergen will extinguish flames from the boxes but will not extinguish the smoldering mass that can re-ignite and start a fire that will not then be extinguished due to the previous discharge of gas. Should we increase the concentration of the Inergen?

A.: Inergen is rated as an ABC agent. The problem expressed by your company manager is one involving deep-seated embers. You would be presented with the same issue if using an agent such as FM-200, and a sufficient "soak time" is what's needed to resolve the presupposed problem on your plate. In other words, the room must be sealed tight. Peace of mind can be reached if the enclosed room can hold the gas for 10 min, in order for the agent to penetrate and extinguish the embers. A securely sealed room provides for expedience of suppression.

Extra agent concentration is not really the answer. For example, FM-200 usually possesses a 7–8% concentration, and if the room is tightly sealed with its door shut, a 6% concentration would completely suppress the fire including the deep-seated embers after several minutes, and definitely within a 10-min duration. Halon 1301 gas does the trick with just a 5% concentration. It's when the gas isn't fully held inside the room that a fire can re-ignite.

Q.: I see no smoke detectors or FM200 heads on the ceiling of our data center, only under the floor. The only thing on the ceiling is preaction sprinkler heads. I have never seen the fire suppression under the floor only and I thought the main idea was to protect the equipment and not have a water discharge. Is it common to have protection like this? Also, won't a discharge blow up and push the floor tiles up and defeat the purpose of the protection?

A.: If you have a preaction system protecting this room, you have dry pipe and believe it or not, some jurisdictions insist on wet pipe for this type of occupancy. It doesn't make sense to me that a contractor installed FM-200 in the subfloor only; why put all your money under the floor? If their mind-set was just to try and save some money, this represents a band-aid approach.

You could have a fire either above or below the subfloor, and there is much more air flow under the floor than most people realize. Remember, the FM-200 cylinders will discharge quite a bit sooner than will the preaction system. They use smoke detectors (spaced to 900 sq ft maximum) for activation, and thus the fire is contained during the incipient stage. The preaction system (ultimately activated by fused fire sprinklers) catches the fire in the heat stage. In your case, what you could do is to use the FM-200 system to release (open) the solenoid on the preaction system to obtain a quicker response for a fire in the room.

The best scenario is to utilize FM-200 for everything, and use preaction as a backup for the data entry room. The controller should be mounted outside the room. Regarding your second question, I am assuming that the FM-200 cylinders are sitting on the subfloor. To allay any concerns, the likelihood is that those raised floor tiles are much too heavy to be pushed up by a below-floor discharge. That would never happen. The primary design considerations should be air flow characteristics, life safety concerns, speed of detection, and anticipating what type of potential fire is to be expected.

Q.: What is the recommended sequence of operation of the kitchen exhaust fan upon fire alarm and Ansul activation? My experience is that the exhaust fan continues to run. Is there any circumstance where the exhaust fan shuts off upon fire alarm or Ansul activation?

A.: After the Ansul system activates, the exhaust fan stays on. What shuts down is the plenum, because you do not want the influx of any make-up air or additional fresh air coming into the fire area. The exhaust fan will carry the chemical up through the bonnet, where grease may be covering the metal shelter above the grill. We want to keep that fan running, and that is why they sticker every hood with a label reading "Do not shut down exhaust fan in case of fire." Of equal importance, energy sources and all power under the hood must be disabled, so that no ignition of any source may still feed the fire.

Q.: I have a physical therapy office in a commercial single-story building. The ceiling height is 9 ft; it is a sprinklered ceiling that also has sprinklers in the roof space. I would like to build privacy partitions using 2 × 4s and 3/8 gypsum board. What is the maximum allowable height I can build the wall without interfering with the sprinkler distribution?

A.: To make certain that you don't have to relocate any existing sprinklers, the highest you can go is about 7 ft 5 in. If there are just a few sprinklers affected, you can reference Figure 8.6.5.2.2 in NFPA #13, where you'll find that if the new partition is right below (or within 6 in. of) the sprinkler, the partition can extend upwards to 3 in. below the sprinkler deflector itself. Or, if the sprinkler is say, 20 in. from the partition, then a partition that is 7ft 10in. high would be alright. But with many sprinklers involved, make sure there is 18 in. of vertical clearance between the deflector of the sprinkler and the top of the short wall, and you will be code-compliant.

Q.: We discovered that when the original fire sprinkler system was installed in our school, sprinklers were installed above and below the drop-ceilings. This was a 1983 installation. Every bit of construction is noncombustible, but for some reason small-orifice upright sprinklers were installed above the ceilings. Was the presence of a tectum roof deck responsible for the decision to do this?

A.: That should not have made any difference. Tectum roof decks have been around for over 50 years. These thick decks combine desirable features such as soundproofing, insulation, and structural integrity. Tectum decks are sometimes installed above swimming pools and coated with a vapor retarder so that they do not absorb moisture. The dense wood fiber is fire-rated, so there is no reason in the scenario you describe to worry about sprinklering above the ceiling. It is absolutely a noncombustible concealed space. To protect against any sprinkler discharging in this space accidentally, you could remove these sprinklers and replace them with 1/2-in. galvanized plugs.

Q.: Our general contractor has proposed that a chain-link fence surround the area in a new warehouse that will contain an electric-driven fire pump and bypass, jockey pump, and controllers. The fire pump will be above grade level. Does code require that the pump be situated in a separate room protected in two-hour rated construction?

A.: What the code requires is found in Chapter 5 of NFPA #20, *Standard for the Installation of Stationary Pumps for Fire Protection*. What is mandated under Paragraph 5.12.1.1 is that indoor fire pump units be physically separated or protected by fire-rated construction. Everyone is so accustomed to the old policy (no fire-rated construction needed), that the new requirement is infrequently adhered to in practice. But if the structure and the pump room itself are protected by fire sprinklers, then the required separation consists of one-hour fire-rated walls. Separating fire pumps from the rest of the building is necessary to provide protection against freezing, floods, fire, explosion, and so forth.

For years the NFPA #20 appendix has included a sentence reading that "some locations or installations may not require a pump house." Back in 1981, the NFPA issued a Formal Interpretation requiring that the fire pump and associated equipment be separated from the remainder of the building with minimum two-hour fire-rated construction. Then, a 1983 Formal Interpretation stated that the fire pump and related equipment need not be separated from all other mechanical equipment and also that "It is not the intent of the Committee to recommend retroactivity of the (1981 Interp) two-hour construction standard unless the authority having jurisdiction determines that the continued use constitutes a distinct hazard to life or adjoining property." However, any actual enforcement of these 1981 and 1983 interpretations seem to have somehow fallen by the wayside in recent years. Until the 2003 edition of NFPA #20, the NFPA had no specific fire-rating policy in this regard.

The real issue here is equipment protection. When an insurance company is highly involved and on their toes, the fire pump will be housed in a separate, heated, noncombustible, sprinklered, and adequately ventilated fire pump room with proper floor drainage provided. This pump room should also be situated as closely as possible to areas where fire sprinkler protection is of the most critical concern. The intent is clear that the fire pump, driver, and controller must be completely protected from mechanical damage originating from any potential adverse conditions, and physically located in their own designated fire pump room.

Q.: We have a small dry system in an unheated portion of our warehouse, with about a 175-gallon capacity. The air compressor servicing the system kicks on and runs about eight or nine times a day. My plant manager says this is no big deal, but is it?

A.: You definitely have a problem. A system that small, if it contains grooved couplings, should be losing less than 1 psi every 24 hours. Your compressor should be kicking in only about once per week. If you told me it kicks on once daily, I wouldn't be overly concerned. But the gaskets in the grooved couplings often dry out over time and lose their ability to maintain a tight seal. Your system should be trip-tested as soon as possible and then tested again annually. This exercise allows you to flood the system, which usually causes the gaskets to seal more effectively.

Q.: Our community uses older model codes as adopted legislation, and our contractor has installed a pull station 5 ft off the floor level, and is being asked to move this lower so that it may be handled easily by the general public including children. Where does it say in the code that 60 in. is too high for pull stations?

A.: The building codes supersede NFPA 72 on this matter. If your village uses the IBC (International Building Code), then you are limited to a pull station elevation that is between 48 and 52 in. above the finished floor. BOCA (Building Officials and Code Administrators; 1993) Section 917.5.1 reads that "[T]he height of the manual fire alarm boxes shall be a minimum of 42 inches and a maximum of 54 inches measured vertically, from the floor level to the activating handle of the box," the same wording would appear in Section 918.5.1 of the 1996 BOCA. This issue is usually addressed in the contract documents by the electrical engineer. BOCA also stipulates that these pull stations must be red. If this is an old job, whoever installed these manual fire alarm boxes will probably still have their name and number located somewhere on the fire alarm control panel. Give them an informative heads-up.

Q.: We have been advised that heated or flaming furniture produces toxic gases. In a fire incident, how lethal are these fire-generated gases?

A.: What is very lethal in any blazing structural fire situation is the prolific production of smoke containing carbon monoxide, cyanide, and other products of combustion, in unventilated areas of a building. At least 60% of all fire deaths, including those of firefighters, result from smoke inhalation rather than burns. Symptoms of smoke inhalation may not show up for 24 to 48 hours after the event. Anyone experiencing symptoms such as vomiting, nausea, prolonged confusion, or difficulty breathing, should be immediately evaluated by a medical professional. The only remedy for the accumulation of toxic gas is ventilation. Venting at the roof is an important strategic tactic, both for the prevention of a possible explosion and also to allow for quick discharge of gases to the outside.

Q.: An inspector flagged us for not having sprinklers beneath some mezzanine flooring that consists entirely of an open metal grate. Can he do this? You can spit through this grating and there are fire sprinklers at the roof which would cover any fire that started beneath the grating on the warehouse floor.

A.: The AHJ (Authority Having Jurisdiction) is there to enforce the code, and should provide you with a code section number to justify his recommendations and comments. If this doesn't happen, reference Chapter 8.5.5.3.1 in the current (2010) edition of NFPA #13. The requirement is noted there for sprinklers to be provided beneath any fixed "open grate flooring … over 4 ft wide." This has been in the code for a long time. The Technical Committee decided long ago on a threshold of 4 ft to determine when ceiling sprinklers alone could not be counted on to control a fire beneath certain obstructions.

Gratings and walkways are included in this section because often these slatted constructions are covered with an ample amount of boxes, random storage, office stock, or floor mats. NFPA #13 also recommends that heat shields be provided with sprinklers beneath open gratings. *FM Global Property Loss Prevention Data Sheet* 2-0, under Section 2.2.1.4, advises to "avoid the installation of open grids because they can obstruct ceiling-level sprinkler discharge. As an alternative, make the mezzanine or walkway solid ..." The data sheet goes on to state that "if open grids cannot be avoided, provide sprinkler protection (beneath the open mezzanine)."

Appendix: A Fire Protection Overview

Without question, we need fire. It is a necessary part of every equation when we cook our food, heat our water, warm our homes, or heat our large commercial shopping malls and high-rise buildings. We use fire to light up cigarettes, to light fireworks for display, and to ignite fuel for lanterns. Nearly every new residential home is built with a fireplace. We use fire to create products, and also to dispose of them. Commercial waste disposal sites are often referred to as incineration plants because they have but three options of what to do with trash: *burn it*, bury it, or recycle it. Those in forestry management often prescribe *controlled burns* for certain remote flat areas of prairie. We need fire, make use of it daily, and it has always been that way.

The 1981 motion picture *Quest for Fire* opens with this narrative: 80,000 years ago, man's survival in a vast unchartered land depended on the possession of fire. For those early humans, fire was an object of great mystery. Since no one had mastered its creation, fire had to be stolen from nature. It had to be kept alive—sheltered from wind and rain, guarded from rival tribes. Fire was a symbol of power and a means of survival. The tribe who possessed fire, possessed life.

Much of the mystery of fire has been solved today. We know that combustion is a rapidly occurring chain reaction in which oxygen combines with some other substance, producing (a release of energy) heat and visible flames. We can quantify and predict at what temperatures certain fires will begin. For any fire to start, we know that there must be three pre-existent entities (fuel, heat, and oxygen) and an ensuing chemical chain reaction. Without one of those, the fire cannot start. Remove one of the four conditions, and the fire is snuffed out. A fire inside a hypothetical "fire-proof" nutshell, for example, will cease to exist as soon as the available oxygen is used up (or the combustible material is consumed). The extinguishing agent CO_2 attacks fire by lowering the oxygen content (our atmosphere is about 18–20% oxygen; the remainder is nitrogen (80%) and various other inert gases). It may take a little while for the application of CO_2 to extinguish a fire, but once the O_2 content falls below 15%, the fire goes out. The application of halon in an enclosed space does not reduce oxygen, but it removes something from the chain reaction that caused the fire to trigger in the first place. Dry powder fire extinguishers operate on the same principle. No one knows exactly how or what it stops in the chain reaction, but it does an efficient job of quickly stopping the fire.[1] So does water, especially in mass quantities, through the reduction of heat. For complete extinguishment, of course, the water must reach the (seat of the fire) solid fuel. That is the only way to inhibit the rate of release of combustible gases and vapors.

Nothing really burns in a solid state; it must first be heated enough so that its surface is converted into a gas or gases, and can then combine with oxygen. Wood is considered a heavy fuel. Its surrounding temperatures must exceed 500°F before the wood will give off enough vapor to sustain ignition. Wood burns at a rate of approximately one inch per hour. Gasoline on the other hand, will burn almost instantly and will continue to burn vigorously. It gives off visible vapors at just about any temperature. A "flash point" is defined as the lowest temperature at which a liquid gives off enough vapor to ignite momentarily in air. The flash point of gasoline is 45° below zero. The flash point of acetone is 3°. The flash point of turpentine is 95°. It is necessary for the heat of the initial flames to sufficiently heat the liquid surface, however, to create a flammable air–vapor mixture in order for flames to spread through that vapor. See Table A.1.

The minimum point at which oxygen will rapidly unite with a fuel is called the ignition temperature of the fuel. Sometimes this is referred to as the kindling temperature. The ignition temperature of sawdust or an oily rag is very low, whereas the ignition temperature of certain combustible metals is very high. Fuel, or matter, exists in three different states: solid, liquid, and gas. Under the right heat conditions, fire is a chemical reaction between the substances oxygen and combustible fuel. Most materials that burn contain hydrocarbons (carbon and hydrogen) which, when burning, produce carbon dioxide and water. The burning hydrocarbons, however, retain almost all the energy, which is primarily released as heat. The heat breaks down more of the (fuel) material, which produces more burnable vapors, and increases heat transfer, making the entire process self-sustaining.

TABLE A.1

Various Flash Points

Liquid	Flash Point (°F)
Butyl acetate	72
Cellosolve solvent	202
Ether	–49
Ethyl acetate	24
Ethylene glycol	232
Fuel oil (types #1 and #2)	100
Gasoline	–45
Glycerin	320
Grain alcohol	55
Kerosene	110
Mineral spirits	85
Styrene	90
Wood alcohol	52

Where there's smoke, there's fire, or at least it's on the way. Most materials begin to smoke before they begin to flame. That's usually because the amount of air in the smoldering or smoking area is in a limited amount. Or, the level of heat has not yet risen to a point high enough to distill off vapors or ignite the various gases that have been produced. Heat can build up in a confined space, leading to an event referred to as spontaneous combustion, defined as "the culmination of a runaway temperature rise in a body of combustible material, which arises as a result of heat generated by some process within the body." The classic example of a pile of wet hay in a barn, or the collected pile of rags (or soiled linens) soaked by some flammable oily liquids are typical examples of this phenomenon. With all the elements for spontaneous ignition, the same scenario is also always present in laundries. The range of minimum autoignition temperatures for textiles varies from 490 to 1060° See Table A.2.

With regard to the actual size of wood, and similar commodities, bigger is better with regard to fire safety. For one thing, the more finely divided the fuel may be, the more surface area there is to burn. And the smaller the volume of fuel is, the lower the volume of heat that is required to reach an ignition temperature. Finally, if arranged willy-nilly in small pieces, there is more oxygen present to fill up intermittent spaces. The cumulative effect of smaller "bundles" of combustible material is a recipe for ignition susceptibility and rapid fire spread. In rare cases, smaller is the way to go. For instance, if

TABLE A.2

Various Ignition Temperatures

Material	Ignition Temperature (°F)
Charcoal dust	356
Corn dust	482
Douglas fir wood chips	500
Isopropyl alcohol	399
Match heads	325
Newspaper	446
Nitrocellulose film	279
Nylon resins	806
Peat	440
Polyethylene plastic	330
Polyurethane foam	824
Powdered sugar	698
Wheat starch	716
Wood	572
Wood fiberboard	430

storing polypropylene plastic pellets (or other small granular plastic objects), a fire will burn less severely because the pellets fall out of their containers and have a smothering effect on the fire while becoming flue space filler.

Naturally occurring hydrocarbon materials were not ordinarily found in everyday life until the late 1950s, with the proliferation of modern plastics. The wide use of plastics in packaging, consumer goods, furniture, and building materials is very nearly an omnipresent threat to life safety. Assuming like weight, the burning of plastics (hydrocarbon based) will produce over twice the heat as will the burning of wood (cellulose based). This is because hydrocarbon-based materials consume 50% more oxygen during combustion over that of cellulose-based materials.

Fire is all about gas. All a flame consists of is burning particles of gas.[2] The rate of combustion is a function of how quickly the chemical reaction of oxidation occurs. All chemical reactions within a simple flame take place in the flame font, a narrow bright cone in which elements are virtually torn apart. Suppose a fire starts in the living room of a house devoid of any fire sprinkler system, while the owners are away. Heated furniture will produce toxic gases that will collect in a vapor cloud at the ceiling. These gas clouds (rich in reaction potential but temporarily low on oxygen) build up heat very quickly until at a point close to 1,200°, flashover will occur. Flashover is defined as the sudden spread of fire over an entire area. In this very dangerous incident, the heat from the gas cloud draws more vapors from the floor and other objects in the room, which violently ignite and burst the whole room into flames. Firefighters receive ongoing education and training to understand the full scope of this lethal aspect of fire behavior.

Fire is our friend only when safely contained. The only problem with fire is that it does not like to be extinguished; it likes to travel around—and grow—as fast as it can. When this country was founded, instances of fire were much less common than they are today. That's because the "controlled" fires weren't as common, at least in the form of heating devices, propane torches, cigarette lighters, matches, and so on. Paper was scarce. Back in those days, no one even knew what electricity was. The reason whaling was such a lucrative industry was because it was whale oil that was needed for lanterns to light up the streets at night. We still live in combustible homes, travel in combustible vehicles, and even our backyards are combustible. The difference today is that we use combustible fuels to power machinery and our cars, cars that use highly flammable rubber tires. We have welding equipment now, and space heaters. Modern life dictates that we maintain proper minimum temperatures to heat our homes. Insurance companies cope daily with the hazards of high-risk industrial occupancies. Hence the need for fire protection engineering and life safety management. Only with the knowledge of the behavior of fire, the threats of hazardous materials, preplanning strategy, system installation fundamentals, and overall preparation for the unexpected, can we comprehensively attack the very real problem of fire potential that will always exist.[4]

Taking Inventory

Conventional wisdom tells us that the present state of fire protection engineering technology is fairly advanced and getting better all the time. But the fact remains that the number of fire sprinkler systems inside structures worldwide, and public knowledge regarding the substantial advantages of fire sprinkler systems, is still in its infancy. There is work to be done for those employed in the fire protection industry and this includes the promotion of fire sprinklers. Fires wreak havoc every day, but one thing history has demonstrated is that in every crisis there is opportunity. So, efforts need to be carefully orchestrated in order to achieve new legislation regarding sprinklers in high-rise buildings, nursing homes, daycare centers, hospitals, new homes, and college dormitories.

This is not to say that the overall efforts of those in the business of fire prevention in recent years have not had a positive impact. Table A.3 and Figure A.1 illustrate the dramatic impact that an increased focus on fire protection and life safety has had on annual U.S. fire losses:

There has been a slight leveling off of the decrease in fire fatalities, relatively speaking, in recent years. Per the National Fire Protection Association (NFPA), the latest fire-related fatality numbers for the United States have totaled 4,030 (in 2003), 4,003 (in 2004), 3,762 (in 2005), 3,334 (in 2006), 3,533 (in 2007), and 3,423 (in 2008). Fires in 2006 accounted for 3,245 civilian fire deaths,[3] a decrease of 12% over 2005. Of that total, 2,580 (just under 80%) fatalities occurred in the home: 425 perishing in apartment fires and 2,155 in typical dwellings. It is alarming to note that, typically, 45% of civilian fire deaths occur in communities having a population below 5,000. Older or otherwise substandard rural housing remains a real 24/7 threat to all who combat fire.

All told, a residential fire occurs every 76 seconds in this country. Ninety percent of fire injuries suffered in structure fires in 2006 occurred in residential properties. Some in society are more vulnerable than others, and that includes children, the handicapped, and the elderly. Also, alcoholics are 10 times more likely than nonalcoholics to die from fire. Half of all fire fatalities result from fires that start between 10:00 p.m. and 6:00 a.m. Fire fatalities peak in the month of January. February ranks second, and December is third. The percentage by which the number of home fires in the United States on Thanksgiving Day exceeds the national average is 32%.

Fire safety education, escape plan reminders, smoke detectors, and a more widespread adoption of residential sprinkler ordinances must be part of the public consciousness in order to generate a further decrease in fire injuries, burns, and fatalities. I was born in the Eisenhower years, when nobody owned a house equipped with fire sprinklers. But no one had a house equipped with smoke detectors either, for that matter, or a car containing seat belts. Now it's the law. What took so long?

TABLE A.3

Fire-Related Fatalities

Year	Fire-Related Fatalites[a]	U.S. Population (in Millions)	Fire Fatality Factor[b]
1978	7,882	221.6	3.56
1979	7,700	224.1	3.44
1980	6,643	226.6	2.93
1981	6,836	228.9	2.99
1982	6,147	231.0	2.66
1983	6,033	233.1	2.59
1984	5,359	235.3	2.28
1985	6,313	237.5	2.66
1986	5,970	239.7	2.49
1987	5,941	241.9	2.46
1988	6,351	244.2	2.60
1989	5,528	246.4	2.24
1990	5,302	248.7	2.13
1991	4,573	251.8	1.82
1992	4,805	254.9	1.89
1993	4,714	258.1	1.83
1994	4,379	261.3	1.68
1995	4,682	264.5	1.77
1996	5,086	267.8	1.90
1997	4,148	271.2	1.53
1998	4,126	274.5	1.50
1999	3,682	278.0	1.32
2000	4,147	281.4	1.47

[a] Includes both civilian and firefighter deaths.
[b] This figure denotes the percentage of 1% per capita fatalities nationwide. For example, in 1998 the fire-related death rate was .015 of 1% of the total U.S. population.

What is it going to take to keep the line on the graph careening downward? One sure way to continue the trend is to tirelessly go after the "big fish." The magnitude of arson activity jumps out when reviewing any fire statistic, and remains an awful menace. Roughly 31,000 structure fires in 2006 were intentionally set, and those resulted in 305 deaths. As grim as this sounds, fatalities resulting from these types of fires averaged 694 annually during the 20-year span from 1977 to 1996. Arson remains the second leading cause of residential fire deaths in this country. Certain cities are more at risk: arson accounted for 46% of fire fatalities in Detroit in 2007, and 25% in Dallas and New York City. Currently, one-fourth of all residential fires in Los Angeles and Cleveland are intentionally set. Only 15% of arson cases are closed by arrest.

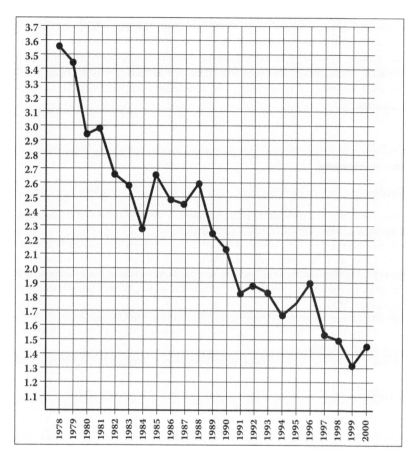

FIGURE A.1
Per capita fire fatality in the United States.

Fires occur daily in this country. Fire is a living, breathing, moving thing. Firefighters know that if they don't hustle to get it, it will come and get them. Their number one priority is to get the wet stuff on the red stuff. For those in the practice of installing fire sprinkler systems, their key to success is establishing the correct priorities: get the design right, make sure all hydraulic calculations are accurate, focus on the potential (worst-case) hazard, and completely sprinkler the buildings. The most important thing for any building owner, particularly one in which large numbers of people congregate, is to actually *have* a fire sprinkler system, ideally a wet-pipe system comprised of steel pipe, properly alarmed, and complete with at least two check valves and a fire department connection.

The fact that there are so many unsprinklered properties in this country is simply unsafe, and the United States continues to have one of the highest per capita fire-related death rates in the world. The cause is not only fire, but also politically driven code management and overall public apathy. Promotion is paramount, and acceptance is necessary.[5] When the general public takes the need for common-sense application of fire protection practices seriously 100% of the time, we are all going to reap the benefits of that amenity. Comprehensive fire safety happens to be something very attainable.

Endnotes

1 No one really knows exactly what happens at every juncture of a pure chemical reaction between 0% and 100% combustion.
2 Although the precise measurement of flame temperature is nearly impossible, we know that the temperature of the continuous flame region (at the fire base) stays relatively constant at 1,650°F. Temperatures drop in the intermittent flame region, as they move toward the visible flame tip, where temperatures hover around 620°F.
3 In 2006, there were also 89 firefighter fatalities in the United States.
4. Mark Bromann, "Our Friend and Foe in a Nutshell," *PM Engineer,* October 2006, pp. 28, 30, 31, 86.
5. Mark Bromann, "Taking Inventory," *Plumbing & Mechanical,* February 2008.

Index

Printed and bound by CPI Group (UK) Ltd, Croydon, CR0 4YY

01/11/2024

01782618-0004